INDIAN ECONOMY UNDER SIEGE

J.D. SETHI

VIKAS PUBLISHING HOUSE PVT LTD

VIKAS PUBLISHING HOUSE PVT LTD
576 Masjid Road, Jangpura, New Delhi - 110 014

Copyright © J.D. Sethi, 1992

HC
435.2
.S4185
1992

All rights reserved. No part of the publication may be reproduced in any form without the written permission of the publishers.

Typeset at PrintCraft, New Delhi 110055
Printed at Ramprintograph, Delhi 110053 (INDIA)

INDIAN ECONOMY UNDER SIEGE

For
Saahill and Pia

Preface

This book was prepared before the new Congress(I) government with Narasimha Rao as Prime Minister and Manmohan Singh as Finance Minister announced the new economic policy. The book had gone to the press and was being serialized in the Financial Express. Subjects covered required some comments on the policy but it would not have made any fundamental difference to the analysis. Therefore, I decided to leave the manuscript unchanged. The main crises that the Indian economy faces today or faced before the new economic policy was announced in response to these crises are still with us. Despite the claims and denials that the new policy meets the demand of the structural adjustments suggested by the IMF and other international institutions or pressures exerted by them, in my view the structural bottlenecks remain intact. It is legitimate to claim that short term measures taken to meet fiscal and balance of payment crises may take us to the next phase when the structural issues would be more conveniently tackled. However, past experience with the various governments has been that because of rigid power structures, even short term, let alone structural, change takes place very slowly. After all it has taken India a decade of liberalization to arrive at a situation where massive new liberalization became inevitable. Besides, one is not sure whether history will not repeat itself because both the balance of payments and fiscal crises became deeper in the last one decade of liberalization.

External liberalization without internal liberalization created many more problems. This point has been sharply made in almost each of the chapter.

Narsimha Rao-Manmohan Singh constitute a team of reasonable men hoping and claiming to avoid financial disasters and bankruptcy which have suddenly burst on India. But the crisis is so serious that only unreasonable men with profound commitment and will can meet these challenges. These leaders are the successors of Nehru-India-Rajiv dynastic, corrupt, opperessive regimes which kept them in cage for too long to allow them to do any thing more than produce conditioned pavalovian reflexes. Their survival instint is as strong as their sense of fear of the coterie. Besides, the orders are also coming from the international financial system.

The new economic policy has been opposed from two opposite sides. One set of economists have called it a total surrender to the international financial institutions and others consider it a half-hearted policy without ensuring implementation. The Government on its parts claims that it has made a sharp break with the past despite homage paid to now defunct Nehru-Mahalonobis model. In my view the new policy is dramatic accentuation of old policies although Manmohan Singh is changing some policies step by step. The rehtoric of self-reliance has been exposed. There is hardly any major industry or defence system which does not have a large external component. Besides there is no system of cooperation among the Indian industrialists and each prefers to have collaboration with foreign companies. The new policy has taken the old to its logical end by removing those controls and regulations which hampered the natural growth of industry, but there is absolutely nothing in the new policy that may bring about the end of the compradore character of Indian business as well as the public sector.

A kind of pavalovian reaction is that whatever IMF-World Bank say or suggest is assumed to be *ipso facto* deterimental to Indian interests and therefore has to be resisted. It is foolish to expect that these agencies are at the service of India. On the other hand it is common place and common knowledge that a creditor must satisfy itself about the solvency of the debtor.

Preface

Some nations have learnt to adjust to the demands of these organisations successfully. Some others just crumbled and became dependent and more crisis-ridden. Therefore, Indian leaders are under severe test. They may deserve support but not unconditionally.

Another response is somewhat whorish. It is argued that when we are outselves introducing structural changes that are required by the IMF, why should the demand be made publicly and cause humiliation. If a whore herself is ready to take off clothes why insist on making a demand. These who now crow about the defence of sovereignity never raised their voices when they seemed to profit from the slow creeping deference to both the USSR and the West. They accepted the slow depreciation of the rupee which was a daily affair. The distinction between what could be yielded and what would be defended is lost in the general rhetoric.

There will be many other responses as policy packages are announced. But India today remains a nation of volcanic convulsions, storms, fires, blood, mini-civil wars, political assassinations and on the edge of economic bankruptcy. There are no short cuts. Solutions are being sought from moment of moment, with each moment bringing India's surrender nearer. The finance minister says he has adopted policies of devaluation and forced liberalization to save India from disaster. This is what Indira Gandhi said in 1966 when the rupee was devalued. Ironically the rupee is being devalued again in a year of less than normal rains. If we face next year even higher rate of inflation and bigger balance of trade deficit, it will not be surprising. But no political leader, bureaucrat or economist has ever been hauled up for this kind of hecloring. The policies of the new government have a limited objective but these are unnecessarily being projected as a matter of great macro restructuring. It renewed not create complacency in us that we have for the first time a largest team at the top.

There are too many forces which have laid siege against India and too many Trojan horses that are preparing to launch new onslaughts to defend their vested interests even if it means generating several mini-civil wars. Who is going to defend India now? No one yet!

I am grateful to the editor of the Financial Express particularly Mr. K.S. Ramachandran to have offered space for serializing the book. Special thanks are to Professor J.S. Grewal for friendly and encouraging cooperation. I also must thank to Library staff of the Indian Institute of Advanced Study for their ready cooperation and to Mr. K. Rehman who typed the manuscript. I am grateful to my wife, Shanti, for bearing with me for all kind of demands that I made on her in putting together the book.

J.D. Sethi

Contents

Introduction: India: The Most Dangerous Decades   *xiii*

1. Indian State and Political Economy since 1947-48   *1*
2. The Long and the Short   *9*
3. Macro Vicious Circle   *17*
4. Fiscal Anarchy   *26*
5. Balance of Payments: Dependency and Deficits   *36*
6. A Vicious Circle of Export and Import Substitution   *47*
7. A Decripit Public Sector   *60*
8. The Ever Rising and Corrupting Middle Class   *72*
9. The Rentier State   *83*
10. Liberalisers Without Liberalisation   *91*
11. Some Structural Issues   *101*
12. Siege by Growth   *115*
13. Towards a Demographic Disaster   *138*
14. A Tale of Two Indias   *147*
15. New Economic Policy   *161*
     Tables   *200*
     Index   *237*

Introduction

India: The Most Dangerous Decades

The 1990s are going to be the most dangerous decade for India. Others said the same for earlier decades and were proved wrong. I hope I shall be proved wrong by those who, through their mystical reflections, have got a blind faith in the nation's destiny.

If the emerging crises are not sternly faced, I fear these might converge on one another, sooner than later, to explode into our face and the prevailing stoic anger and inertia may burst into violent fury. India may not survive the next two decades if her power elite, instead of preparing itself to meet the multiple challenges and crises remains wilfully engaged in the search for a disaster. Indeed, the time is running out on us.

There never was a period over the past 44 years when India did not face one crisis or another—sometimes of a very serious nature. But never before had she to contend with the convergence of such serious crises, both internal and external, as she faces now. It is tragic that the elite response is either collective acquiescence, breast-beating, or panic or resort to down-right lies. It seems that there is no scope left for truth or honesty in India's political leaders. Indeed, we have become a nation of liars. Pick up any newspaper in the morning and reach the truth about the lies anywhere and everywhere. But as stated in the Preface, the crisis is so serious that only unreasonable men with profound commitment can meet these challenges. Nevertheless, we cannot get a better than had Manmohan team under the post-dynastic dispensation.

Volcanic Convulsions

India faces four major threats, all at the same time. First, the Indian political system is dead and cannot be revived. Second, the Indian economy has been pushed into a situation of premature decay and galloping dependency. Third, social violence is reaching such dangerous proportions as may rip the society apart both horizontally and vertically, and, finally, the massive intervention by all major powers in the areas around India creating an unprecedented threat from outside. Externally, India is being marginalised and yet her economy is under siege.

The demise of the political system took place some years ago. But the politicians in their innocence believe it not an act like a mother who, refuses to believe the death of her child. Political scientists act like maggots entering this body through various wounds. The death of the political system has orphaned political functionaries.

We may have had a series of elections but with each election the political institutions have been disrupted, corrupted and now destroyed. Political parties have become irrelevant because to win an election the political machine is needed no more. The judiciary was undermined, humiliated and its powers have been usurped. The bureaucracy was nearly paralysed but it responded by building an internal empire that was corrupt and stalled growth. The educational system, except that part of it which was appropriated by a small and affluent minority, has totally collapsed. Not only have the institutions been destroyed, but all those norms and values which imposed discipline on political elements have been thrown to the winds.

No Exit From Politics

Yet there is no exit from politics. It is a remarkable aspect of Indian life that once a politician, always a politician, whether in power, in legislature, in committees or completely outside power.

Introduction

Most Indian politicians and so called political activists cannot make an honest living outside politics. According to one estimate, there are about two million such perpetual parasites feeding on the system through means and measures which could not but wreck any system. The Indian economy and society are too weak and poor to retain this vast non-functional army with the limited economic surplus it produces.

The death of the Indian political system is being debated but with such irrelevancies as the choice between the parliamentary and the presidential form of government, the basic and the non-basic structure of the Constitution and the Centre and the States. It is a great reflection of the bankruptcy of the Indian Marxists that they have reduced every aspect of Marxism to Centre-State relations only or, externally, to anti-Americanism.

In fact, what has been thrown up is a loose combination of vastly expanded authoritarian and anarchic forces, so loose as to be ineffective and yet so intertwined as not to allow a new system and new institutions to grow and function. The line between the elective and the non-elective systems has disappeared. Attempts being made to revive, artificially, the dead political system, instead of giving birth to a new one, make no sense.

Elite Leadership

There seems hardly any possibility for the rise of a new political system until such time as a new leadership emerges. The entire second generation leadership of both the ruling party and the Opposition has turned out to be corrupt, incompetent and devoid of all political norms. The attempts made to create a third generation elite leadership by the Naxalites, the RSS, the Janata Dal, finally, the Mandalites and Kamadees, have all flopped. Indeed, some of the roughnecks of the defunct third generation leadership have proved so nasty and brutish that it seems that the alternative to the present system would be some kind of a dictatorship.

Poverty

The most startling aspect of the situation is the premature decay of the economy. It is quite surprising that whereas for over a decade and a half the rate of gross savings and investment has moved from about 12 per cent to about 22 per cent, the rate of growth has declined from about 4.5 per cent per annum to about 3 per cent in 15 years from 1965 to 1980. But when it was pushed up to 5 per cent during the eighties, it collapsed. The use was illogical thus, it is not quite surprising that, consequent upon this development, the number of people who have been pushed below the poverty line has been increasing. The rate of growth of money-supply, inflation, budget deficits and the balance of trade deficits, have all doubled or trebled everyday. It is difficult to think of another country in the world which has got into this kind of a mess. There is no economic theory to explain all this.

It is for all of us to note that the economy is totally mismanaged, both in the private and the public sectors. The infrastructure, such as power and transport, has declined in performance so much as to make nonsense of any further planning. Not to speak of earning profit, public sector undertakings are unable to provide even for interest on borrowed capital. Industrial sickness is spreading like an epidemic. Misallocation of resources has destroyed economic rationality. The rural-urban dichotomies and the respective lobbies that lie behind them are making the system increasing rigid.

The crisis of industry is its growing sickness and low growth rate, both of which emanate from sources that have exhausted their capacity to generate and stimulate new growth. Besides, tariff protection, subsidised finance, underpriced infrastructure and intermediate goods supplied by the public sector, have all combined to bring about a premature decay of Indian industries.

The most important reason for industrial sickness is the inter-elite conflict as well as the joint elite effort to maximise their illegal gains by a variety of methods. The fiscal and regulatory system has been such as to give the bureaucracy a big share in graft, and to allow business to manipulate production, distribution, gross earnings and profits. How much of production is illegally sold is anybody's guess. But the fact that when even conditions of high demand, availability of credit and raw materials and moderate taxation are met, firm after firm keeps adding unlicensed and yet apparently "unlicensed and unutilised" capacity.

This is a fraud unknown in history. The crux of the problem is that when the owners can make larger profits by illegal means than otherwise, they develop a vested interest in sickness because once an industry is declared sick, numerous other concessions from Government can be utilised for reducing costs in addition to what is taken out from unlicensed capacities.

India entered the phase of technological stagnation and even decay some years ago. This is not so obvious to many; even slight visits to industrial complexes give the impression of tremendous progress. This progress however relates to the past. Some new technologies are also being added every year, but the rate of technological process is slow because of continuous reliance on imported technologies not being supported by the creation of a system of linkages between import and adoption, innovation and research and development. It is yet to be established whether new liberalisation policies benefit technology competitiveness.

How does one measure technical progress? One macro measure is increase in productivity. In industry and manufacturing, productivity has declined sharply. The rate of growth of productivity now is less than one-half of one per cent per annum and this is in sharp contrast to about four per cent in the fifties, to about two per cent in the sixties and over 1-1/2 per cent in the first half of the seventies.

The situation on the labour front is far more serious in another way. It reflects stagnation or decline in labour productivity. In this field, the role of the public sector has been disastrous; no norms of efficiency are observed by anyone, be they administrators, technicians or workers. In general, India loses 30 to 40 million mandays every year and no labour legislation, however stiff, can meet the challenge because there is no serious cost consciousness among decision makers at all levels. And now this figure is reaching the 50 million mark. On the other hand, the power of the trade unions has reached a point at which it has become totally anti-work. Labour has become the enemy of productivity.

The most serious problems of the Indian economy and polity

is the massive generation of black money. It is true that black money is generated in every country, but the important difference between India and other countries is that in India black money has even crossed the limit at which it becomes inconsistent with the efficiency of the system. In recent years, black money generation has been growing at a very fast rate in this country, so much so that it has become difficult to isolate a single sector of the economy or administration which moves without the use of black money. In fact, black money does not oil the system any more; it obstructs decision-making. Most economists are unanimous in their view that black money generation has become a major factor for inflaming inflation, because it not only misallocates resources but also shifts them from investment to consumption. Above all, black money generation makes nonsense of all anti-poverty programmes because, in the final analysis, this money comes from the pockets of the poor.

Social Relief

India has faced many political and economic crises in the past but never did she face the kind of social crisis it faces now. Earlier, the severity of conflicts and violence was confined to communal problems. Over the past few years, social violence has spread over many fields. Indeed, social violence in one field is now setting up a chain link by its extension to other fields. The most disturbing aspect of the situation is the general belief that violence is the only course left for defending oneself. It is possible to think of law and order measures, education and social work when conflict, is, by and large, social. But when social and economic conflicts coincide, the result often is large-scale bloodshed. The new characteristics of Indian politics are (a) mini civil wars, caste-isation, violent conflictualition all over, criminalisation and institutional distruction. What is common among all is towal alienation and legitimacy of violence.

The alienation of the masses from the elite and the authority is also complete. There is no respect left for law and order, and the law openly appears as an instrument of oppression and violence.

Introduction

In view of the complexity and the mutually contradictory nature of the social crisis, it cannot transform itself into a revolutionary movement. It often creates new outlets for violence while the status quo remains.

The most ugly aspect of social violence is the participation of the police itself in criminal activities. At no time in our history has the police been hated so much as it is today. Third degree methods are used by the police to extort information leading often to deaths in prison. If social violence, reinforced by police violence, is not checked, it will overtake both the economy and the polity. Indeed, it has already overtaken a large part of both. The very survival of the Indian society is now at stake.

As a result the nation is getting divided against itself everywhere. There are two Indias in more than one sense and they are moving on a collision course: (a) those below the poverty line and those above it; (b) the relatively developing eight States and the stagnant rest; (c) areas that are becoming susceptible to permanent militant intervention and the rest; (d) the minority of the English-language educated elite and the rest; (e) one set of communal and caste clusters against the others. Rural-urban dichotomies and rigidities have reached a stage at which economic policies relating to both have become impossible to pursue and which are pushing urban and rural elite into a collision.

Authoritarianism

We have left behind practically all the optimism, assumptions and beliefs about what Independence was to bring. Every belief now seems to have been grossly misconceived. The emotions stirred by the rising tide of nationalism have become feelings of bitterness. Democracy may or may not have had much significance for the bottom half of the population, but the survival of democratic institutions had been regarded more confidently by the other half. Now that hope has gradually disappeared, if not the actual death of democracy, at least into the fear of its future extinction. Politics is in the malevolent grip of those who have no respect for democratic norms and, instead, have the guts to treat people with contempt. The scale of hopes unfulfilled was seen dramatically in the 1980s, and is now seen in the rising tide of the new and old forces of corrupt elites.

I do not believe there ever was a dearth of solutions to our problems. If, despite the solutions all our problems survive, it is because we refuse to see what is happening in the black boxes of Indian politics and to speculate about the future. Whether optimistic or pessimistic, speculation about the future is a risky intellectual pursuit, though we need not be as squeamish as William Wilberforce was when he wrote in 1790: "I dare not marry; the future is so dark and uncertain."

Truly, the current crisis is not national but civilisational; India is not merely a territorial entity or a nation in the normal sense of the word: it is a civilisation which has already been sliced two or three times. It would be utterly futile to make comparisons of India's crisis with that of any other country which does not have a civilisational character. Leaders come and go, systems rise and fall, crises multiply and disappear, but it takes as long a time for a civilisation to decay as it takes to build. The Indian civilisation, which has stood the test of many centuries, onslaughts, corruption, internal crisis and sabotage, now seems to have entered the state of massive decay. Unless this process is drastically arrested, India will go the way of other civilisations.

The incompetence of the leadership that caused the demise of our political system, the dishonesty and corruption of businessmen, the cynicism of the bureaucracy, the massive confrontation of social forces, etc., are only part of the process of this civilisational decay. The real cause of worry is not the crisis but the incapacity of the power elite and the intellectuals to stand the psychological strains produced by this decay, the dis-integration of the fabric of civilisation, and the massive human alienation.

It is not necessary to follow any particular philosopher's scheme of rise and fall of civilisations, but we must learn to respect and reckon with the negative and positive historical forces operating in that direction. Most European observers had regarded the social systems of the eternal East as models of tradition that were incapable of modernising themselves. Mao Tse-tung proved them

Introduction

all wrong. He rejected the linear path of the West. By drawing inspiration both from his own past and the present, he was able to safeguard the Chinese tradition and civilisation from decay by modernising them on the strength of their own values as much as on new knowledge.

In the first instance, Mao looked upon China not as a nation but as a civilisation. The Chinese civilisation had also declined and decayed for many centuries. Mao was faced with the problem of integrating the civilisational problem with the problem of building a nation-state. Indeed, his task was even more difficult because he faced another challenge—of international Communism which at that time had subjugated itself to defending the Soviet national interest. Stalin was dead opposed to the Maoist line. Mao successfully fought Stalin by fixing both nationalism and Communism. Mao's remarkable achievement was that he succeeded in combining the rejuvenation of the Chinese civilisation with the building of the Chinese nation-state and the Marxist ideology, something which was unheard of before or ever since. The failures of Mao Tse-tung, such as the Great Leap Forward and the Cultural Revolution were no failures. In fact, these were unnecessary surgical operations and blood letting to complete the comprehensive civilisational revolution that Mao wanted to accomplish.

Political Sense

In our own country, Mahatma Gandhi faced the same problem when he came on the Indian political scene in the beginning of the century. In many respects, the situation of decay and pessimism of today is not much different from what prevailed then. The socio-economic stagnation and the cultural decay of those days had pushed the Indian political leaders and intellectuals towards accepting the cultural superiority of the West in every respect. Mahatma Gandhi used an unusually violent language to denounce the then Indian elite for its collaborationist and subservient attitude towards the Western civilisation.

The situation then corresponded to what Marx described as despotic and static or what Orteg Y. Gassel referred to as "anthropomorphic vegetations". Toynbee called it "animalism" characteristic of arrested civilisations, i.e., societies in which the pattern of behaviour was like those of lower animals on pre-determined lines. Toynbee described the static situation of decaying civilisation thus: "The equilibrium of forces in their life is so exact that all their energies are absorbed in the efforts of maintaining the position they have attained already, and there is no margin of energy left over for reconnoitring the course of the road ahead, the face of the cliff above them, with a view to further advance."

Mahatma Gandhi was the first modern Indian to have addressed himself to this problem and made civilisation the unit of analysis at a time when imperialism reigned supreme. Against the imperialistic nature of the global system, he successfully combined the struggle for creating consciousness about the Indian civilisation with the struggle for making India an independent State. He never gave up this task although his colleagues and successors could not fully appreciate his views and conventionally betrayed him.

He made the struggle of Indian civilisation against the West and against its own conformist elite the focal point of the national movement. He mobilised the Indian masses and used the mass line to isolate those Indians whom the British had weaned away. His small book "Hind Swaraj", though crudely written was a seminal work in the sense that (a) it reiterated the superiority of the Indian civilisation over others and reformulated its values; (b) it exposed the hollowness of those who denigrated the Indian civilisation; and (c) it gave bare out-lines of the kind of future social political and economic structure that would be consistent with the value system of the Indian civilisation. Quite remarkably, he though accepting non-violence as a creed and a way of life, forcefully stated: "I will not hesitate to use any amount of violence if my civilisation was threatened".

Indian Marxists may like it or not, but both history and Mahatma Gandhi proved that Marx was utterly wrong in his prediction, because even as he ignored the civilisational aspect of the change in India, Marx had predicted that British imperialism would play a double, positive as well as negative, role in India. On the one hand, it will destroy what he called "ancient static communities of India" and on the other, create a new Western type of civilisation. Almost as a vengeance, India has produced hybrid culture and, in retrospect, it seems that Mahatma Gandhi was right when he called Brtish imperialism an unmitigated disaster.

What Mahatma Gandhi saw was the danger of British imperialism leaving behind enough of its cultural transmitters who, in the name of science and rationality, would destroy precisely the basic fabric of the Indian society. India was not a small country to be transformed in a short time into a society of the Western type. Mahatma Gandhi chose Pandit Nehru partly because the latter had earlier talked in civilisational terms and partly because Mahatma thought Nehru alone would be able to integrate the warring factions inside the Congress, while others may divide it.

The mistake that the post-war leaders made was that they politicised everything and put every problem of the Indian civilisation into the political perspective of a nation-State. They also reestablished old hierarchies which had been the main cause of the decay of the Indian civilisation. They reintroduced the idea of a new caste and status in politics and society. Nobody is more conscious of the status today than the Indian intellectuals. Through, status and hierarchy, the Indian culture and intellectual life got bifurcated, partly outside the context of the Indian civilisation and partly inside it, thereby developing spilt personalities over the entire leadership spectrum.

Nowhere is this more sharply reflected than in the two per cent English-knowing educated classes refusing to see the communication problem of the rest of the 98 per cent of the population for which English makes no sense. Mahatma Gandhi gave the highest priority to this issue of communication between the masses and through his unflinching effort made the entire country accept simple Hindustani as the *lingua franca*. His successors failed singularly to carry forward this tradition and if there are not one but two Indias today in the cultural sense, it is because of that mistake.

It is the middle class which always carries the civilisation values. Pandit Nehru created a large middle class in the hope that it would act as a buffer between the State and the masses. But, he paid scant attention to two aspects of this class: (a) the numerical growth of the middle class has been several times the growth of the working class—something which was bound to create an economic crisis; and (b) that this class could become the enemy of the Indian civilisation by refusing even to look at it, not to speak of enriching it. In one sense, the crisis of our society is the crisis of this parasitic class because the economy cannot produce enough economic surplus to feed its fancies and dysfunctions. The class is completely alienated from the masses.

The collapse of norms and values of our society and the middle classes is a direct expression of this civilisational crisis. The society is sick to the point of death. In the face of colossal human suffering and poverty we pride ourselves on being cheerfully cynical. Our universities and research institutions are no longer places of learning and scholarship; they are merely waysides and their influence on the ordinary man's already diminished integrity is so fierce that the latter inflicts his own inadequacy on the nearest and dearest. The value-setters, social reformers and revolutionaries of today are unfamiliar sorcerers. India is being ruined by the contagion of despair about the neglect of its civilisational values.

Sometimes, people express amazement at the loss of values and the recrudescence of such ugly phenomena as communal violence. There are not the expressions of a transitional society moving into a new one. Instead, they are the symptoms of the decay of societal values.

There are tides among nations when powers rise and fall. The states grow stronger or weaker. There may be riptides among nations, vast terrents of change in politics, and economics and culture sweeping away old structures and creating new ones. In this tide Mahatma Gandhi represented the collective conscience of the Indian people and their civilisation and his successors the collective interests of the elite. India is not merely facing economic or military competition with other nations. She is pitted against civilisational forces whose strength is now formidable. The external threats to the country need not take the form of external aggression.

Introduction

It is enough that her social and cultural fabric is destroyed by all kinds of open or settled intervention through culture, technology, ideology and now IMF intervention and India ruling elite's willingness to surrender. The ruling elite which is alienated from its own people is tied, one way or another, to the economic and military and cultural interests of the advanced nations, is now ready for absolute surrender and betrayal. What happened two hundred years ago is repeating itself as a much bigger tragedy. The new surrender and betrayal are taking place with tremendous intellectual dishonesty fortified in the name of defending India's economic interests and strengthening the economy at a time when India along with many other countries has already been declared an international ghetto. In the final analysis, all old and new loans have to be repaid.

Conclusions

The problem in fact is no longer only economic, it is a problem of political economy. The term political economy is defined an economic system which openly accepts the constraints of political power structure which itself is sustained by a certain kind of distribution of economic wealth and income. All developing countries are known for high inequalities of income and wealth and it is impossible to think of a political power which would be ready to destroy this economic configuration. However, countries which have a democratic set up find that democracy itself requires a different kind of economic power distribution, notwithstanding all the manipulations that the ruling elite may be indulging in. In other words, a norm-based democracy is a necessary condition for any macro strategy or model of development. But this is not a sufficient condition.

The sufficient condition which is consistent with the aforementioned four strategies is the one in which the State identifies itself with the interests of all the producing classes and is ready to crush parasitic classes. The Srate does not need to have any particular ideology other than the ideology of identification of the interests of the ruling political elite with national interests.

National interest is defined as interest articulation and integration which while guaranteeing the basic needs and jobs allow the power elite freedom but also demands obligations to fulfil the stated objectives.

It is better to know this arrangement by means of examples. For instance, Japan or South Korea and Germany on the one hand and China on the other have produced different political economy models in which the state identifies itself with the interests of the nation and the producing classes and provide basic needs to every one. They, thus, indirectly ensure full employment. In terms of economies their models are different but not in terms of political economy as defined here. China is a communist nation. Japan is a fullfledged democracy. South Korea and Taiwan are half way between dictatorship and democracy. But they all have one thing in common, namely, identification of the state with the interest of the nation in a way in which gross inequalities, poverty or exploitation are reduced, if not removed, through the dominance of the producing classes.

CHAPTER I

Indian State and Political Economy since 1947-48

There is now a general agreement that neither economics can be discussed without politics nor politics without economics. There has to be a political economy approach for the analysis of Indian economy. But what is political economy? Karl Marx described the works of Adam Smith, David Recordo and Malthus as theories of political economy. There was nothing specifically political about their respective works and theories. Simply because each of them so entitled their main work as exercises in political economy was not a good enough reason to call them political economists. In terms of theory, Karl Marx was the first political economist and Schumpter was the last one. Keynes fell in between. Mahatma Gandhi defined his approach as Moral Political Economy. For him, every problem was at once existential as well as moral. The tragedy and force of Indian economists lay not only in their wholesale borrowing of western theories but also in their fettisoning of the moral dimension of every theory. The present crisis is as much moral as it is economic. The siege makers are not only economic parasites but also moral degenerates.

By political economy, I mean (a) nature of the state, (b) linkages between economic and political and the power structure, (c) the political and economic process as they flow from one another, (d) the amount of economic surplus the political system takes away. In the case of India, an additional component is (e) the massive generation of black money in the system effecting both the political economic decision-making.

India became independent on August 1947. She inherited a very difficult situation created by partition. Million of people died, wounded or were thrown out of their homes. The irony is that Pandit Jawahar Lal Nehru said that he accepted partition for the sake of economic development. History has passed a harsh judgement on that facile assumption. It is even more strange that Pandit Nehru accepted the vivisection of India almost in defiance of Mahatma Gandhi's vision and the next day he called for tryst with destiny. The destiny was the return of old imperialism in new forms. He left the Kashmir problem unresolved and later opted for a defence system against China that bordered on total unpreparedness. We are still living with the consequences of second-rate men and women playing tricks with the Indian civilisation and destiny.

The strangest of all was, as we shall discuss later, that after having driven out every radical element out of the Congress, Pandit Nehru tried to impose on the party and country a socialistic pattern of development. How could a right wing party be an instrument of socialism?

Imperial Bureaucratic State

The most important development in the post independent period was the creation of an Imperial Bureaucratic State (IBS). Instead of destroying the old constitution—the old imperial constitution of 1935—almost the whole of it was incorporated into a new one. The main difference between the two was the introduction of the adult franchise in the latter. At the same time, Pandit Nehru entered into an alliance with the ICS which was the most solid pillar of British Imperialism. The question as to how a class of civil servants who till yesterday was serving imperialism would be the instrument of national transformation the next day was never asked. Once the Imperial Bureaucratic State was set up everything flowed from it as the logic of that system. Pandit Nehru kept on creating the illusion that India was on the way to a great transformation. Sometimes people say that he gradually introduced transformation. People also argued that Pandit Nehru had good intentions and he had a grand vision but it was the bad implementation of his policies that resulted in the distortion and crisis. All these are untenable arguments. The man who set up a bureaucratic state should not have expected anything different. In fact, Pandit Nehru had no vision of any kind. If he had one, then he openly surrendered it to a class of bureaucratic consipirators. As a chosen successor of Mahatma Gandhi and a darling of people he let both of them down by his enormous intellectual deception. The people did not appear in his model.

At best, the state set up under the new constitution, followed subsequently by the setting up of a Planning Commission, could have been meant only for a Liberal-Fabian society and not an egalitarian society. The founding fathers of the Indian Constitution were great legal luminaries, but they were sociologically a poor lot. There was only one economist in the Constituent Assembly, who, however, was brushed aside. The founding fathers paid no attention to the ugly but dormant massive underlying social conflicts which were suppressed for centuries and were ready to come into the open at any time. For such a situation, India needed a radical constitution and a mobilising state. There was a great faith in Pandit Nehru as a socialist and a mobiliser. But he was a liberal fabian of aristocratic origin. He certainly could not rise above his class. He betrayed the masses who gave him whatever he asked for.

Four Revolutions

Pandit Nehru went even a step further and created the impression that under his leadership, India would launch all the four revolutions of our time: (1) Nationalist revolution (2) Deomcratic revolution (3) Industrial-Technological revolution and (4) Egalitarian or Socialist revolution. The Indian intellectuals, who almost entirely were either Marxists or Western liberals hailed this impossible ideology of four revolutions. It is difficult to imagine any important country opting for that and not facing a crisis on all four fronts.

The setting up of the Planning Commission created a stir among the intellectuals, as if that institution would solve all of

India's economic ills. Although the Planning Commission has served some useful purpose, its main role was to provide the economic counterpart of the imperial bureaucratic state. India was proclaimed as a federal polity but the States were not given real powers as guaranteed by the Constitution. One of the reasons why the Planning Commission was not made a statutory body was that in case this was done similar statutory bodies had to be set up in the States. Over many issues, Centre-State relations reached a boiling point for which people have been seeking an explanation. The real explanation lay in the setting up of the Imperial Bureaucratic State with a centralised Planning Commission which was not backed by planning from below.

Something had to be done to give the impression to the people that they have also been involved. Three institutions without teeth were set up for this purpose. These were (a) Panchayat Raj system, (b) Cooperatives and (c) Community Development. It is no accident that all the three institutions were smothered or became dysfunctional because no constitutional protection was provided for them.

The Indian Constitution concentrated only on the functions and powers between Centre and the State. It rejected any further Constitutional devolution to the third or fourth tier of the Government. It is again an irony that the man who fought against this devolution was Dr. Ambedkar who feared that the Village Panchayat would further strengthen the landed, trading and moneyed interest against the landless and the under-privileged who mostly belonged to the schedule castes. Obviously, he had no faith in the democratic process at the bottom. Today, we find that in the absence of a Constitutional guarantee for election at the local level, the power structure in alliance with the local and state bureaucracies oppress the rural poor in ways which were unknown before.

The Imperial Bureaucratic State was not merely an internal arrangement. Had it been so it might have been transformed. It was also linked with the old colonial system. The 1949 policy resolution which ensured equality between Indian and foreign capital despite stiff opposition from Indian industrialists was an

Indian State and Political Economy since 1947-48

act of absolute economic treachery. It laid the basis for economic dependence. At the end of the war, the British capitalists who dominated Indian industry were selling their businesses to Indians. But by one stroke of the pen, Pandit Nehru reversed the process and obliged the British capital.

A delegation of Indian industrialists led by Mr. Purshottam Dass Thakur Das met Pandit Nehru and advised him against the 1949 policy. Pandit Nehru refused to budge and, in fact, insulted them. Therefore, in the first few years of Independence, the Constitution and the institutions set up under it created (a) an Imperial Bureaucratic state; (b) the Planning Commission for centralised planning; (c) Three very weak basic institutions of Panchayat, Cooperatives and Community Development, thus leaving out the masses in decision making except for voting in the general election and; (d) Linking the Indian economy back with the former imperial economy.

Mahalanobis Plan

In the early 50's a new debate was started which ultimately culminated in the formulation of the Mahalanobis Plan and which was hailed as the most radical plan outside communist countries. Ironically, the Planning Commission was manned by the old ICS. It was too much to expect of a conservative Commission to initiate a radical plan. Nevertheless the plan was seriously debated. It thrilled economists as it created the impression of putting India on the path to rapid industrialisation with the public sector having the commanding heights in the economy.

The Mahalanobis plan nearly completed the structure of Indian political economy in which the imperial constitution, concentration of power in the hands of ICS, a centralized Planning Commission and above all the public sector controlled by the bureaucrats and not the professionals formed the core components. This was still not a complete model. What was left in it was completed by the imposition of massive regulations, controls and licensing. The last component unleashed inter-elite and intra-elite conflicts which corrupted the whole structure.

In terms of economic calculus, the Mahalanobis Plan suffered from several shortcomings and contradictions. First, when actual calculations were made it was found that employment and investment targets were wide apart. Second, the Plan made very small allocations for agriculture and yet expected it to produce the surplus both for the Government and the market. Third, the consumption goods which were to be produced by labour intensive sectors were subjected to policies which favoured the capital-intensive sector. The Mahalanobis Plan was ushered in without a consistent policy frame. Fourth, the scale of foreign aid, foreign capital and import of technologies were not made explicit. Last though not the least, social planning and objectives remained divorced from economic objectives.

It is not possible to go into other detailed criticisms of Mahalanobis Plan but some of the distortions, dichotomies, contradictions, and stratifications it produced have to be exposed. It is a great pity that, the Indian Marxists and the other Left oriented intellectuals fell victim to the dominant classes and a corrupt political elite. The Left made the mistake of equating socialism with public sector, nationalisation, state intervention of controls and regulation. The result was that the economic objectives were pushed aside and the Imperial Bureaucratic State came to terms with the big business, big landlords and most of all the rentier class. Indeed, the Mahalanobis Plan generated the most preferred class, the rentier class. This character of the Indian state and planning let loose a conspiracy between the political, bureaucratic and economic elite.

This conspiracy did not merely deprive the millions of their livelihood by concentrating economic activity in small enclaves but was also the cause of rampant corruption, inefficiency and anti-democratic, anti-federal policies. The consequences of this conspiracy was not only disastrous for the economy but also for the polity.

Continuous centralisation of power in the hands of authorities in New Delhi, large-scale violation of norms and internal imbalances within the centralised power elite thwarted solutions to these problems. The kind and degree of manipulation that the political elite indulged in to undermine both the judiciary and Parliament, with the help of the bureaucracy, also upset the horizontal balance of power between the executive, Parliament and the judiciary. Unfortunately, within the executive itself power was concentrated only in the hands of the Prime Minister, so much so that many analysts have dubbed Indian democracy as a "prime ministerial dictatorship"

Four separate but linked problems have emerged from the political economy, functioning of the Constitution and the values and behaviour of political parties and the power elite over the last four decades which bear on the whole polity. These are (a) tensions and imbalances between the Centre and the States; (b) shift of power from the national to the regional parties; (c) wide inter-regional disparities; and (d) crises of planning on account of lack of devolution of functions and power below the State level. In addition, the corruption, the communalisation, criminalisation and lumpenisation of Indian politics, further complicated these problems.

A Sense of Desperation

A sense of desperation instead of cool calculated analysis has overtaken the system. Consequently, institutions of polity and democracy are now becoming an arena of no-holds-barred political battles in which the instrumentality of the state along with money and muscle power increasingly determine the outcome of the next elections. The political leadership seemed to believe that once the electoral battle was over, adjustments in crisis management would become the normal state of affairs. This belief was dangerous and unwarranted after repeated defeats.

In a federation which functions on a party system, the rise of regional politics and regional parties is inevitable. If the federation also passes through a phase of one-party hegemony which generally subsumes regional politics, regional political forces are bound to come on the surface with the end of this dominance. Indian polity has witnessed both phases. The struggle between centralising and decentralising forces inevitably produces a balance or imbalance between regional and national parties, depending upon which force is more powerful. However, so long as both the Centre and the States stick to norms and laws and function within the same Constitutional, legal and political system, there are limits to the threat regionalism can pose to the Centre. But if norms are violated, imbalances and conflicts are bound to emerge.

The phenomenal growth of regional ethnic sub-regional parties —there are scores of political parties registered with the Election Commission of which only eight are multi-state parties—has rendered many of the assumptions about the Indian polity largely infructuous. The quasi-federal Constitution which was designed as a political expedient to foster unity in a society which encourged diversity and which was elaborately structured to provide countervailing forces against excessive federalising and contrifugal tendencies, has not proved neutral with regard to the relative shift of power between regional and national parties. One-party hegemony, which was expected to act as a powerful extra-constitutional resolvent of Centre-States, inter-party and pressure-group conflicts, has been followed by the most dangerous party fragmentation. But most dangerously, it created a situation of mini civil wars and terrorism all over.

Economic development through centralised planning, a growing public sector and fiscal predominance of the Union, was also designed to set functional limits on the growth of regional political power. The predominance of these centralising economic tendencies has ended now, leaving behind intolerable pressures for decentralisation, a prolonged phase of economic stagnation and, inevitably, almost the end of old style planning itself.

Given the crisis of the Centre-States tensions on the one hand and massive violence and subversion, actual or potential, in several States on the other, it is important that any national party or coalition ruling at the Centre must neither try to suppress the regional sentiments nor allow its power to be eroded. Power must be Constitutionally devolved to lower units of the state if the balances in the polity and economy are to be restored. The Imperial Bureaucratic State must be dismantled through the decentralization of power down to the villages if Indian political economy and the state are to remain stable and ensure development with justice.

Chapter 2

The Long and The Short

It is rather unfortunate that the battle lines are being drawn between the Government and the opposition and between the economists of different pursuasions on the post-economic policies of the Government. It seems that the original optimism about creating a consensus was no more than political rhetoric. A tragic-comic aspect of the battle is that both sides still remain protagonists of the now failed Nehru-Mahalonobis model. Most of the disputants, at one time or the other, contributed to the making of the prevaiting crisis. However, all Nehruvians of today are lame ducks and are superficially debating solutions that defy the problem. What makes matters worse is that they themselves are part of the problem. Policies being pursued or proposed in respect of economic imbalances, instead of removing them, may plunge the economy into a tailspin. Even if we get, at the end of the year, a new pattern or package of polcieis, one is not sure of its success because the implementation still remains in the hands of insti-tutionalised bureaucratic structure.

The Government is claiming that it is set on the road to introduce structural changes. It is also true that without structural changes economy cannot be taken out of the deep and multiple crisis it has run into. However, there are sharp differences over what constitutes a structural crisis and which structural problem. One man's construction may be another's de-construction, to use a post-modernist metaphor.

The Government is not expected to reveal the distinction between steps taken under international pressure and those taken independently to correct past mistakes. But the economists and the opposition should remain alert to the distinction. Some decisions taken under pressure may have been unavoidable but others must be resisted. What is of criticial importance is that new domestic policies should not be adopted under pressure for old or new vested interests and parasitic classes some of which are tied to the international system.

It is difficult to capture the current economic crisis within the framework of any one of received doctrines by which the Indian economists or planners have moulded their thinking and policy-makers their practice. In fact, even the combined analytical force of several models based on these doctrines will fail to explain entirely the Indian economic crisis, let alone yield a way out. The two main reasons for this bottleneck are: (a) the terminal decay of what was in the first instance, an impermissible and impractical transplantation of imported models and (b) a total intellectual disruption that surprisingly has overtaken the Indian mind. If some economists are still stubbornly debating dead ideological issues which concern only a privileged few, the exercise cannot merely be denounced as escapism, it is downright intellectual bankruptcy.

The Indian economists are agreed upon the fact that the crisis is very deep and structural and, therefore, is not amenable to short-term policies. One has to remind oneself that most of the economic theory and controversies surrounding five year Plans have revolved around the short-term analysis.

The analysis for the next decade requires two kinds of exercises. The first is the dissection and delienation of the existing situation and its likely projections for the foreseeable future. The second is the kind of environmental changes, both national and international, which impinge upon the development of the economy and society.

To me, the Indian case appears to be an ideal empirical reference for all the major crisis theories. Every crisis theory finds its validation in the prevailing situation. We need not go into these theories but anyone having a theory can easily see its test in the Indian empirical conditions. Briefly in respect of the economy, the Indian crisis is marked by the co-existence of *Short-term, Cyclical, Structural and General* (pathological) crisis. Correspondingly, it is a multiple challenge of *Decay, Recovery, Growth, and Adjustment*. A chronic pathological state will not respond to political or administrative mediation from those who are both the perpetrators and the victims of this crisis. The challenges of the 90's, therefore, are compounded and thus permit of no compartmentalised solution. Nor do they permit the facile assumption that mere theoretical or analytical existence of a right kind of solution will automatically discover leaders who will accept and operationalise that solution.

Development Model

Let us first pinpoint the real parenthood of the Indian development model because its children are still disputing about it. Stripped of historical disputes and normative considerations, it can be shown that there is no paradigmatic/difference, say, between Ricardo, Marx, Keynes and Friedman. It is now agreed that almost all the economic theories of the last 200 years, evolved since the days of Adam Smith are based on a single, unique and powerful paradigm. All the controversies and disputes between various schools of thought have been, and remain within the framework of that unique paradigm. The paradigm was falsely imposed on India. Therefore, it is no accident that the ideological battle lines drawn among the Indian economists have all been exposed as belonging to the same imperial genre. If the old issues are being debated in the same old manner, this unrealism is peculiarly Indian and acts as a barrier against sharper articulation about the prevailing crisis.

Though most economists do accept that the crisis is structural, they do not take the trouble of explaining the meaning of the word structural. This word has different meanings in different social science disciplines. For instance, structuralism and its internal critique post-structuralism, are theories of positivistic political philosophy which has been most fashionable in the West in recent years. Political scientists had talked about structuralism much earlier denuding politics of values. Economists, except the Marxists, had avoided the use of this word and, instead, coined their own phrases and concepts to explain away the same kind of phenomenon. Such analytical theories as trade cycle, long waves, equilibrium and dis-equilibrium, recession and depression, long-term and short-term are some of the well-known ways in which a structural problem has either been explained or cleverly concealed.

One can analyse these concepts and phrases but it is important to be more precise about the meaning of the structural crisis in India. A structural economic analysis by definition, assumes that the components or variables in the structure are mutually interacting, being linked backward and forward, producing trickling up and down effects at the non-agrarian structure, the stronger these relations are.

One of the ways in which an economic crisis is defined as structural is that the economy is caught in a series of vicious circles. This is my approach. It is worth reminding ourselves that four or five decades ago when the development economics and economic development were jointly pushed into the centre of the economic debate, it all began with a spotlight on the vicious circle of underdevelopment, *Poverty, low savings, low capital formations, low growth rate and back to poverty* moving in a vicious circle and these were structurally related. Nurkse, Lewis, Kalecki, Robinson and our Mahalonobis were all familiar historical figures who analysed economic development in structural terms. India quite unnecessarily fell into the Western trap and focused on breaking this circle by aiming at higher rates of savings to achieve higher capital formation and growth while assuming that, in the course of development, poverty and inequities will take care of themselves. Even these Western paradigms contained such components as technological development, employment, agriculture and industry balances. However, these were taken as derived or of second order importance with disastrous consequences.

Everybody agreed that we have had to squeeze planning time not only to catch up but to solve age-old neglected problems, if a violent political revolution was to be averted. The remedy lay in substituting it by more constructive developmental revolutions. Broadly, six transformations have been involved in development and one or the other was suggested to developing countries as a course of their respective economies to follow: (1) Industrial Revolution; (2) Agricultural; (3) Transport; (4) Technological; (5) Institutional and (6) Change in values and attitudes. The four of these were mentioned earlier. Modern developed nations have laid substantial emphasis in all these directions and one could escape fulfilling the demands of each of these revolutions. But the central issue was as to which was the main motive force or initially developed and was followed by others. Since no nation had had all their revolutions at the same time, the spread effect and the trickle down, implicitly or explicitly; became substitutes for revolution.

Unfortunately, we delayed or turned away from these revolutions. We have moved into a situation where several vicious circles have gripped us at the same time, ironically as a consequence of breaking the classical vicious circle of underdevelopment. They are mutually reinforcing. Thus, the task of economic analysis, economic planning and breaking the vicious circles has become very difficult. Unfortunately, economists and planners have been looking for solutions within the *status quo* without changing which no vicious circle can be broken. There is a tendency to analyse the prevailing economic crisis outside the structural framework, but such efforts can only result in the paralysis of analysis. It is completely futile either to rely on old fashioned analysis or to search for short-term palliatives. Even those who are engaged in day-to-day economic activity, such as businessmen or farmers or policymakers, are profoundly inclined to get out of the mess in the shortest possible time or route and with quick fix policies, if possible, without affecting the structure.

Another difficulty in clearer economic thinking about planning and development has been the assumption that Five Year Plans could be formulated as if no crisis of major dimensions would overtake the Plans within a five year period. This was the most absurd assumption as Plan after Plan was overtaken by a series of crises. There has been no crisis-free Five Year Plan or Plan without crisis-planning. Ironically, attempts to meet one crisis often generated two more. Remarkably, no Five Year Plan saw its outcome not being very different from what was aimed at the time of its formation and yet this scheme remained the same. Therefore, it is necessary that while analysing the structural crisis of Indian economy, one must carefully build into analysis the answers to the other kind of crisis of which we have had a lot of experience.

Since India ignored the real developmental revolutions, now she is facing not merely economic but also many more serious political and social upheavals, mini civil wars and mutinies. Therefore, no economist can seriously claim to give a meaningful analysis outside the framework of political economy and a certain value system both of which were excluded from the traditional economic theory. Mahatma Gandhi had repeatedly warned Indian intellectuals and leaders against falling into a trap of economism and scientism which were the products of Enlightenment and Renaissance phase of European history. In other words, economic analysis has to spread its net wider than it is permitted by the narrow economic theories. Thus, we have to get out of the clutches of the received doctrines and unless we do that, we would not be able to explain the working of the various vicious circles.

No Clear Distinction

This crisis also comes about by our not making a clear distinction between objectives and policies, on the one hand and policies and programmes on the other. The elements in each pair overlapped to blur the distinctions. But, the clearer the distinction say between policies and policy goals (objectives), one can locate, more easily, the cause of the unsatisfactory outcome. There has been another order of problems—the objectives are generally mutually conflicting as are policies. But, if objectives cancel one another because of the inherent conflicts, then the planners face brickbats on some easily identifiable objectives, and in our case, it had been the growth rate. This reductionism has let to people calling into question the very exercise of planning as it failed even on the final outcome of growth.

The conflict of the economic structure with the political power structure as well as its internal conflicts was so sharply brought to bear on the translation of policies into programmes, their implementation and the control and effectiveness of the delivery system that the Planning Commission became everybody's whipping boy. The explanatory variables of the power structure, the parameters of a rigid social system and the control mechanisms of implementation did not allow much to trickle down and thus exposed acute and frequent disparities between objectives and outcomes.

There was another trap too. Economists being votaries of behaviourism and *positivism* draw facile conclusions from trend statistics. What is the meaning of the trend statistics in this situation? The trend statistics have a meaning in the sense they average out all kinds of positive and negative influences and cyclical *oscillations* and bring out more realistic co-relations between variables. However, it must be said that, in our kind of situation, trend cannot be either extrapolated or projected in the future easily because the averaging may lead to different results in the future. No matter how much the trend figures are nearest to reality and reliable instruments for current analysis, it amounts to perpetuating the *status quo* and structural rigidities and to using the results derived from them for proposing transformation.

Finally, I do not know whether it is a tragedy or farce or both that the Soviet model i.e. Fel'dman's model, which was used by Mahalonobis to produce the Second plan model was more successfully adopted by other countries with a more committed leadership than the Indians had. Prof. Hirishi wrote in 1976 at a time when the Indian model had slumped and Emergency clamped: "In fact it can be said that Japan has literally followed the development path indicated by the Soviet growth model-referred to as Fel' dman's model by Domar—in which the proportion of investment devoted to producer and investment goods industries is the strategic factor in accelerating the rate of growht."

It was a model not of socialism but of efficient capitalism which the Indian Marxists vehemently advocated as well as sabotaged. I do not know whether contemporary planning is within or without that model. But it seems the vicious circles, which were our own creation, have laid a seige against the Indian economy.

CHAPTER 3

Macro Vicious Circle

There are several structural distortions or interlocking vicious circles in which the Indian economy has been caught. These have to be broken at once if the economy is to be retrieved from the protracted and deepening crisis. In other words, it may mean that we have to search for an alternative strategy or model of development.

The first vicious circle is between *growth, employment, poverty, and inflation.* Our development strategy as embodied in the Nehru-Mahalonobis Second Five Year Plan model ostensibly aimed at accelerating the pace of savings-investment rate and industrial production, with capital goods sectors awarded to the public sector more or less on a monopoly basis. The share of these sectors was more than half of the total public sector Plan investment. The neglect of agriculture which produced an early crisis, distortions and slowdown even in industry were corrected a decade later. However, the nature of this partially corrected development and the distribution of its burden were so designed as to keep nearly half the people in poverty, whether they were inside or outside the development process. Notwithstanding many claims towards having reduced the poverty levels, it seems that no serious dent has been made on the problems of age-old poverty and deprivation, particularly if non-caloric measures are also included. Indeed, in many ways, the situation has deteriorated. Besides, growing population has washed away many benefits which could have gone to the people. The development pattern itself was no less responsible for the increase in population, thus adding a new point to the vicious circle.

Most of the benefits of development have gone to a small percentage of people, mostly to business classes quite naturally, and to the middle classes not so naturally. At best, the upper and middle classes taken together constitute about 20 per cent of the population and take away about two-thirds to three-fourths of income and wealth of the country, leaving the majority in poverty and destitution. This is also the class which is responsible for the highest contribution to savings. In fact, as development unfolded itself and since the public sector made negative net savings, it was this class which, in fact, financed a large part of the fiscal deficit and public sector capital investment. The Government was the borrower and the middle and upper classes, euphemistically called the *household sector*, the savers. At the same time, it is also this class which not merely consumed to its satisfaction but also flaunted vulgar consumption at levels sometimes not even known in the developed countries. Thus, a nexus developed between this class, which held a complete sway over the political power, and the state by which this process of appropriation from income along with high rate of saving and consumption were made possible. What upset the model or the circle of privileges is that the public sector which took about half the investment resources stands in yields in inverse relations to investment made into it. Although the ratio of Government expenditure to national income which went up to high as 40 per cent, the Government sector as a whole contributed practically nothing to net savings in the economy. Over the last few years, both the Government expenditure and the public sector investments have been financed through very large and accelerated deficit financing. Consequently, we have the following outcome:

(a) A modest growth rate, (b) fairly high rate of savings and investment, (c) a moderate rate of inflation of less than 10 per cent until very recently, (d) fully entrenched poverty of about half the population, (e) the most vulgar consumption of the upper and middle classes which maintained high demand with high level of poverty; (f) widening inequalities.

Poverty, A Barrier

How does the first vicious circle move? One can begin with any of the aforementioned aspects. Let's begin with inflation which moves both ways on the circle. Inflation in India compared to any other third world country is rather modest but it is large enough for transferring incomes from the poor to non-poor so that both the size and the structure of poverty remains intact and the growth of goods of mass consumption is kept low as income shifts to saving classes. On the other hand, if a large majority of the people are poor, which, by definition, implies a low level of demand. Unless the economy is totally mismanaged, there are good chances of inflation being contained. Add to it, a high rate of savings and consumption confined to the top 20 per cent of the population. The marginal and average propensity to save of this class never goes above one-quarter to one-third of its income. Thus, the Indian poverty remains a barrier both against high inflation and structural radical social transformation, let alone revolution. People have low incomes and, therefore, low demand and lower pressure on the market. People are busy from morning till evening to eke out their miserable livelihood that they are incapable of staging a social revolt. In desperation, they may resort to sporadic violence, but cannot stage a revolution which requires sustained efforts. More out of guilt than a serious response, the Government opted for anti-poverty programmes which are designed to create an illusion of social and economic mitigation. Therefore, a part of the first vicious circle has the following relations: a limited growth rate which does not trickle down, a large proportion of the national income is appropriated by a small class which is enough to both save and indulge in vulgar consumption; mass consumption remains low and as expected the growth of consumer durables has been twice to thrice that of the other consumer goods. This keeps inflation at a moderate level and thus perpetuates poverty. Corruption and inefficiency in the public sector retards growth. High population growth rate ensures the perpetuation of poverty. It is in this way a part of the circle gets its dynamics.

We cannot go into the details of the characteristic of each variable or relationship but an illustration will suggest the complexity of the problems. For instance, what are the characteristics of Indian inflation which was a cause as well as instrument of maintaining the vicious circle. Apart from remaining comparatively low compared to that in other developing countries, reflecting poverty and social inertia, it has been accelerating though moderately from one phase to other, rising from less than 5 per cent in the first two decades to about 8 per cent in the subsequent two decades; its effects are more quickly transmitted through the narrowing of the gap between price changes and subsequent adjustments by producers and consumers; it absorbs a large proportion of external component through high input prices such as oil, capital goods, metals etc; it retards export promotion; and finally, it is responsible for creating domestic distortions of investments and resource allocations.

On the other part of the circle move three variables, namely, investment, growth and employment, along with the other three i.e. savings, inflation and poverty, they make the circle complete.

The trend growth rate had been around 3.5 per cent until the beginning of the 80s. It is to be noted that as the saving and investment ratio increased from 50s to the 70s, the growth rate did not go up appreciably. In fact, in the first decade and a half it reached the peak of 4.5 per cent and declined to 3 per cent in the second decade and a half. The growth rate got accelerated in the 80s but so did the costs. There is considerable dispute and confusion not only about its statistical verity but also about the unaccounted and unstated costs such as heavy external and internal borrowings, environmental degradation and decline in employment growth. But there is no dispute about the fact that now higher growth rate of the economy is possible with higher prices and higher deficits.

Two Lacunae

When the Nehru-Mahalanobis plan was formulated, economists had pointed out two serious lacunae in it, apart from the fact that such a plan could not have been implemented through the bureaucracy in a situation of adult franchise and social rigidities. Implementation needed mass mobilisation. However, in the first instance, the plan had deliberately neglected agriculture and investment in the rural areas and assumed the Stalinist type of possibility of extracting surpluses from agriculture for investment in industries. It was not realised that such an approach would create a serious bottleneck in terms of wage goods. This difficulty in terms of food supplies was unexpectedly made up by enormous quantities of food made available by the United States under PL 480 law. It was an unexpected bonanza and prop for plan. Not only was imported food made available as a wage-good, it also provided enormous counterpart rupee funds for budgetary support for investment in the public sector. The group of economists supporting the Nehru-Mahalanobis model were jubilant and attributed the early growth to the model itself to the powerful force of the model itself and not to the external props. Yet the model's drumbeaters abused the U.S. for its food aid.

The second lacuna in the model was that employment growth was kept outside the model and was expected to go up as a derived consequence of investment, notwithstanding the fact that more than half the investment in the public sector and about the same in the private sector was to go in the capital-intensive industries. Employment growth in this sector averaged about 1.5 per cent and in the last few years slumped to about 0.6 per cent as against 6 to 7 per cent per annum in East Asian countries. In general, for the economy as a whole employment growth remained around 2 per cent and never went up irrespective of the level of investments. In the meantime, the population growth also revealed an upward trends, thus creating the vast army of the unemployed and the underemployed. The Planning Commission could not face this enormously complex problem and thereby, even the calculations of employment and unemployment from the Third Plan onwards. At this time came the NSS into the picture and started producing figures of unemployment and under-employment. The NSS figures provided the proverbial fig leaf for the underlying structural failure of the development strategy in respect of employment.

It was only during the 80s when the growth rate went up to about 5 per cent per annum and about which the economists went hysterical that the figures of employment revealed the ugly picture. Almost in all sectors except services the growth rate of employment declined throughout the 80s. The relationship between poverty and unemployment which had been ignored came suddenly on the surface providing one more proof that there was something structurally wrong with the Mahalonobis model of development.

The other two variables on the circle namely, savings and investment did not show any significant change during the 80s after having peaked at the end of the 70s. However, the structure of savings continued to change in favour of the middle and the upper classes, even more sharply as the fiscal deficits increased. Indeed, the saving rate could only be maintained by strengthening these classes, who were both the savers as well as the large consumers of luxury goods. What was not explained was why the saving rate did not go up, though it was quite clear that the decline in the savings in the public sector and the corporate sectors was signalling towards savings-investments stagnation. During the 80s, the Government resorted to heavy external borrowings revealing and reinforcing the gap between the domestic savings and investments was widened.

If one-third of the income generated is black, how does one calculate the GDP? If the larger channels are turning black money into white, what meaning can one give to the relationship of the saving and the GDP? What qualitative value can one attach to the high rate of savings, if the whole basis of such savings is the result of the perpetuation of poverty more or less at the same level. It is obvious that high poverty, low inflation, high savings, black income, immodest growth rate, all move in a circular trap. Had India's growth rate been not around 4 per cent but double that rate, we would have a very high rate of inflation other things remaining the same. But, the vicious circle would have been broken and surely there would have been some serious dent on poverty and unemployment.

Therefore, as can be seen from the circle, the vicious circle of relationship between the six variables was made complete. Whatever the reasons for fiscal and balance of payment deficits they simply reinforced the response of one variable to the other. The grand macro-vicious circle has taken hold of the planners and policymakers and they have established enough vested interest to maintain the circle. The only snag is that the people of India are now violently responding to the Government inertia and demanding a break in the vicious circle. Whether it would be violence or a rational change in strategies that would ultimately break the circle remains to be seen.

With the economy caught in too many monkey wrentches, the economists wasted time and effort on one simple issue, whether or not there was a genuine acceleration of growth rate during the eighties as against the trend rate of less than 3.5 per cent maintained since 1951-52. When the fiscal and balance-of-payments crises exploded at our faces, economists were silent on the issue of growth.

Assuming that there has been some acceleration of growth during the eighties, we have to answer several questions if we want to visualise, as we must, about the future trends in the nineties. In the last two years, the growth rate has declined, but this may be due to short-term reasons of political instability and the Gulf war.

We need not take seriously the official claims which must inevitably paint a rosy picture of the economy as the achievement of higher growth rate by the Government of the day on its supporters. Late Prof. Sukhomoy Chakravarty once noted that the ICOR "appears to have come down somewhat in the course of the last two five-year Plans since the annual average rate of growth of GDR has been around 5 per cent over the prior 1975-85". No proof is offered. Every thing 'appears' i.e. it is best inferred. The data for different sectors have not been collected. Obviously, if the rate of investment remained changed, the higher growth rate must inevitably be due to decline in ICOR. This is a growth identity, not an economic analysis. Now that the figures are available, they tell the opposite story. There has been no downward trend in ICOR. Indeed throughout the eighties, it remained above 5:1.

Another explanation is there. During the last 15 years, there has been a dramatic shift in the structure of GDP from commodity production (both agriculture and manufacturing) in favour of the service sector. It is well-established that the capital invested in the service sector has a much lower ICOR. Indeed, if the sector with lower ICOR expands at a faster rate than the sector with higher ICOR, the overall ICOR is bound to remain the same or even decline. One conclusion is very clear. ICOR is of no use in explaining any of the current crises.

Anyway, the 5 per cent growth rate is far lower than that of the successful economies which have crashed out of the under-development trap. What is the big deal? Dr. Mitra has drawn attention to this fact: "There is, therefore, a seemingly disproportionality in the recent shift in the composition of India's national income. The explosion in the service sector cannot be readily attributed to any impulse transmitted by the sector engaged in material production."

The Real Cost of Growth

The most important question is not the nominal but the real cost of the growth which does not get reflected in GDP figures. At least two or three costs are of such profound significance that even a purely statistical economist cannot afford to ignore them. There are fiscal as well as physical costs particularly massive environmental degradation, about 100 million people left handicapped and, of course, the perpetuation of mass poverty. Besides, there were monetary costs and the external debt and acceleration of inflation rate which produced the inevitable result: higher the growth rate, the more misery and unequal distribution. Inflation dried up all the trickle down effects.

We cannot go into details but it is quite obvious now that the macro vicious circle can be broken only at the point of employment now, which implies the rejection of Nehru-Mahalanobis model.

Macro Vicious Circle

In all the five year Plans of the past, employment at the macro level was treated either as a residual or was derived from growth strategy. Even at the time when the Mahalonobis plan began it was shown that the plan did not have any internal consistency in terms of relations between sectoral growth and employment. We need not go deep into history but the fact remains that over four decades, whereas the growth rate more or less remained constant, employment elasticity having gone up first, sharply declined later, particularly over the last decade. Surprisingly, in some sectors it became negative in the eighties when the growth rate was higher.

CHAPTER 4

Fiscal Anarchy

The second vicious circle is derived from the first one and it is confined to the fiscal system. At the time of Second Plan, it was decided that nearly half of total investment would be made in the public sector. Although, it was assumed that as the public sector expanded more and more, resources would be internally generated in the sector and after a while the draft of the public sector on the budgetary support and the rest of the saving pools would disappear. This has not happened. The crisis and the paradox was that as the public investment continued to usurp half the plan resources, the budgetary resources also increased. At the same time, the Government's consumption expenditure went on increasing at a very rapid rate for all kinds of reasons that originally did not figure in the Nehru-Mahalonobis model. If today, 70 per cent of the Central Government revenue expenditure goes to meet the obligation of debt services, direct subsidies and defence and the large chunk of the remaining goes to meet the salaries of the bureaucracy, it is not difficult to conclude that the Indian state is run by a small class by and for itself, i.e. for and by the upper and middle classes who preferred as much of conspicuous investment as of conspicuous consumption. The larger the investment in the public sector, the smaller and even negative became the return on investment and still larger became the total subsidies reaching 45 per cent of the budgets of the Central and State Governments, the burden of which fell on the poor. This was a totally irrational fiscal policy from all ideological viewpoints and this could not continue because the burden on the poor could not be increased beyond a point and also because the ruling class was not prepared to tax itself beyond a level which was inconsistent with its high rate of consumption and savings. Therefore, increasingly the Government relied on fiscal deficits which now seems to be getting out of hands, i.e. about nine per cent of the GDP and 200 per cent of the normal revenues.

An important aspect of the Government of India's fiscal arrangement had been that, it in order to achieve a high rate of savings on which the Government's budgetary borrowing programmes depended, many tax concessions and shelters were created. Consequently, the Government fiscal system came to depend more on borrowing than on taxes, as the high marginal rate of taxation, which was hiked to as much as 97.5 per cent, proved counter-productive.

The Direct Taxes Enquiry Committee wrote "When the marginal rate of taxation is as high as 97.75 per cent, the net profit on concealment can be as high as 4300 per cent of the after-tax means", i.e. "it is more profitable at a certain level of income to evade a tax of Rs. 30 than to earn honestly Rs. 1000".

The more Government borrowed, the larger became the burden of debt servicing. Since there was reluctance to catch the tax evaders of the middle and the upper classes, the Government resorted to borrowing even for the purposes of paying of old loans.

All this was symptomatic of the ultimate limit of Mrs. Indira Gandhi's populism. When the economy slid into a lower growth rate left economists rode on her handwagon. But when the tax rates were reduced, no attention was paid to trimming the expenditure incurred on populist programmes. Besides, the evaders had tasted blood. The law enforcing agencies joined in the conspiracy to defraud the Government. No one in India has ever got a jail sentence from tax evasion.

Economists have failed to notice that it was Mrs. Gandhi who changed the gear of the income side of the Government by shifting from taxation to borrowing. It was an enormously successful shift but it proved self-defeating for two reasons. It was embedded in the rapid expansion of middle classes. It was also accompanied by a rapid increase of corruption in expenditure and a further decline in internally generated revenues from the public sector.

The tax structure itself left out a large chunk of taxable income either on account of tax avoidance or tax evasion. The Government failed to expand the tax base and, in times of crisis, relied not on pushing up the rates of taxation but on borrowing, so much so that ultimately the Government was obliged to borrow to service old debts.

Instead of the generation of black money shrinking it was further accelerated because of the new nexus that developed between the business, politicians and bureaucrats. This was the direct result of a vast system of controls and regulations imposed by the Government in the name of the plan but actually for strengthening the ruling class. The political power elite maintained their internal power balances by squeezing the society. In the process, they also generated more black income and nearly wrecked the economy, creating the most subversive vicious circle.

Professional Class

The professional class which expanded at a fast rate and constituted the bulk of the middle class found the system convenient for tax evasion. Lawyers, doctors, chartered accountants, architects, engineers all fattened on the system and resorted to massive tax-dodging apart from benefiting from the tax shelter provided by the Government. The crux of the matter is that the vast expansion of the middle class took place by the use and perversion of the fiscal system which was modified repeatedly to enlarge its savings base.

Therefore, the second vicious circle acts in the following fashion, increasing Government expenditure and public investment make a large draft on private savings and incomes; the inefficiency of the public sector and its incapacity to produce an internal surplus obliges it to depend on the budgetary resources which require a further Government draft on the private incomes; private incomes are not taxed beyond a point because the ruling class cannot hurt itself; the Government helps this class increase its savings from which the Government borrows instead of taxing it; the ratio of direct taxes of Government incomes declines, the indirect taxes contributing as much as 85 per cent of the Government's revenue though now there is a limit beyond which they cannot be increased for fear of inflation and, therefore, ultimately the Government resorts to deficit financing. This is the second vicious circle in which the whole fiscal system is caught.

Changes in the interest rate policies increased the burden of internal and external debt and directly and indirectly strengthened the sources i.e., the middle and upper classes. The 80s also saw an upward shift in the entire interest rate structure such as deposit rate, private sector bond rate, and most significantly the Government lending and short-term borrowing rates. The Government borrowing which forms a substantial part of the Government income becomes costly with the rise in the interest rate. During the first three decades, the interest rate was kept deliberately low in order to finance the public sector. This was made possible by pre-empting bank deposits through the stagnating liquidity. The real rate of interest was thus kept below the equilibrium rate which forced down or caused stagnation in the growth of savings. Unfortunately, the rate of interest was increased at a time when the growth rate of both public and private investment had already stagnated. Which was the cause and which was the effect is not clear but the fact that the domestic resources stagnated shows that the interest rate did not have much influence on household savings; it simply increased the costs and strengthend the middle class.

However, the burden of the public debt went on mounting as a result of the high interest rate so much so that we have reached the position where additional borrowings have become inevitable in order to service the old borrowings. Consequently, borrowings for captial expenditure gradually declined. In fact, both the revenue and capital expenditure are now financed by the printing presses. In other words, during the eighties there was financial liberalisation on the one hand and double fiscal expansion on the other as against earlier financial repression through a low rate of interest that was artificially kept down and limited deficits. The period of fiscal repression produced a high degree of fiscal indiscipline. Thus the eighties, which saw a rise in the interest rate, financial liberalisation and fiscal deficits, also saw a period of strict credit rationing and control of banks. This is a new contradiction that has enlarged the fiscal vicious circle.

A Critical Aspect

A critical aspect of the vicious circle is dynamic relations between budget deficit, money supply and inflation. The objective may have been to mobilise larger resources for the Government and the public sector but the methods used resulted in firming up the vicious circle. For instance, the gross fiscal deficit of the Government rose from 3.5 per cent of the GDP in 1970-71 to nearly 9 per cent in 1991. This had led to the acceleration of the creation of the money supply which induced higher inflation. Higher inflation, in turn, induced larger budgetary deficits as the Government expenditure was kept at an unreasonably high rate. Consequently, a higher rate of money supply was created leading to further inflation. The monetarist economists have had a field day. They had a clear view of their theory, namely, that the money supply was the culprit and the rest followed. Therefore, they upheld a view that the vicious circle started from the money supply. The Keynesians and post-Keynesians have attributed inflation to changes in Government expenditure. The two Western schools of thought fought their mock battle in India. Now, we know that both theories can be easily married and we had the worst scenario of both reinforcing each with little time lag.

In other words, one can determine the vicious circle from any point and end it wherever one likes, money supply or Government expenditure with different results. The causation of variables run both ways and have involved variables which are both independent and dependent. The perception of administration and the power elite is limited to the single issue, namely, whether prices change more quickly in response to changes in money supply or changes in expenditure. Although on this aspect rested the theoretical divide between the monetarist and others, and econometric models were used to prove one or the other hypothesis, the distinction for all practical purposes was of secondary importance. However, the general approval has been to control credit to the private sector, sometimes to absurd limits. The Reserve Bank has been guilty of showing no resistance to expansion of wasteful Government expenditure. Consequently, it has undermined many productive sectors and nearly wrecked the banking system.

What came out of econometric studies in India strongly revealed the existence of the vicious circle between *budget deficit, money supply and prices*. There are other points in circle, whether independent or dependent. For instance, the dynamics of three variables was the expansion of domestic and external debts as the Government demand for borrowing and interest rates went top. Increase in the expenditure of the Government further widened the budget deficit. To the extent that the Government expenditure is productive, higher revenues can be legitimately collected. But, if the revenue expenditure increases faster than the capital expenditure or the latter is inefficiently and suboptimally utilised, the result would only be bigger deficits and that is what happened precisely. This suggests that one should bring into the analysis the variables of productivity and economic growth. The vicious circle became more vicious because of low rate of economic growth and its internal structure increased pressures on prices and government expenditure. Thus, as shown by Narender Jadhav and Balwant Singh through the econometric model of fiscal-monetary-growth-prices, this vicious circle or dynamic nexus yielded estimation results which were quite devastating in the sense that "Each and every parametre have the expected sign and almost all co-efficience are statistically significant. Each collection has explanatory power (as measured by R^2) more than 90 per cent... The similar results indicate that the model almost out-performance other variable models".[1]

Fiscal and Monetary Variables

It is not surprising that at the face of it, all crisis and vicious circles revolve around fiscal and monetary variables, generating

---

[1] E.P.W. January 20, 1990.

the superficial view that everything would be all right only if we somehow could solve our monetary and fiscal problems. But, these problems are symptoms and reflections of the real forces, Akmal Hussain is right in saying that "The mechanism of economic growth in South Asian countries has been such that while poverty increases, the crisis in economic structure manifests itself increasingly in the form of financial crisis (E.P.W. August 18, 1990).

The four important financial variables which attracted attention are (1) Inflation, (2) Budget deficit, (3) Balance-of-payment deficits, (4) Debt-service ratio and debt accumulation.

We can take a circuitous route and probably through this arrive at the same conclusion as arrived otherwise. Also, one method can be checked with the other. One vicious circle can be explained in terms of the other. Some of the crude financial and non-financial indicators have revealed a disturbing picture.

First, the rate of inflation has accelerated with the increase in growth rate during the eighties. Indeed, whereas the growth rate of the economy has gone up by one per cent point and a decade, inflation has gone up by five per centage points. Second, the budget deficit as a percentage of total Government revenue has gone up.

Third, the growth rate has been up during the eighties but its real cost is too high to sustain it at that level and without damage. Fourth, the share of services or non-commodity production has gone up against commodity production laying the basis for structural inflation. Fifth, the officially calculated rate of savings after having reached the figure of 20 per cent over two decades has remained there since then. Sixth, poverty remains fully and deeply entrenched although the Government and the Planning Commission claim that the poverty ratio has declined. Judged against other life sustaining variables like health, sanitation, consumption of proteins and vitamins, drinking water facilities, there has been no significant improvement. Indeed, in some respects, there has been deterioration. Seventh, employment elasticity has uniformally gone down in almost all sectors except two. In other words, the period of relatively higher economic growth rate has been accompanied by a decline in employment growth rate. Eighth, not finding a satisfactory explanation, the growth economists attribute growth to a decline in ICOR which is an identity and not analytically proved. A shift of GDP in favour of services could have caused a decline in ICOR but the corresponding increase in capital and material intensity of commodity production confirm that a higher rate of growth cannot be sustained with the present structure of production or of GDP. The capital intensity of manufacture has increased from 41 to 51 per cent. Ninth, data on income distribution show that the lowest twenty per cent have suffered an income loss, the highest two dicile have increased their share.

Tenth, although the growth of the Government expenditure as a proportion to national income declined during the eighties both in real and normal terms, the ratio remained still so high—one to three times that of the growth of the national income—that the budget deficits kept on mounting and will keep mounting for years to come if the growth in the tax revenue does not increase. However, the most significant and rather unproductive aspect of this decline is seen not in the non-Plan expenditure but in the Plan expenditure. The non-Plan expenditure as a percentage of total expenditure has increased from 55 per cent to 68 per cent whereas the ratio of the planned expenditure to total expenditure had declined from 43 per cent to 32 per cent during the eighties. At the same time, the direct subsidies have increased from 12 to 17 per cent of the budget. Most of the subsidies have gone to the non-poor.

The paper on Financial Dimensions and Sectoral Allocations in the Eighth Five Year Plan (1990-95) approved by the Janata Planning Commission at its meeting on the 18th September 1990 had highlighted the massive problems of resource mobilisation.

The paper had drawn attention to the seriousness of the internal fiscal crisis facing the country. It stressed that the proposed public sector plan outlay (Rs. 335,000 crores at 1989-90 prices) could be implemented under conditions of reasonable price stability only if the declining trend in public savings was reversed by a combination of measures to raise the Government revenue to

GDP ratio (from 19 per cent in 1989-90 to 22.5 per cent in 1994-95; substantially moderate the growth of Government consumption and food expenditure and subsidies to release Rs. 50,000 crores for the Plan; and improve public enterprises profits.

When Indian planning began Keynesianism and Left-Keynesianism were hot stuff. One of the guileless and utopian outgrowth of Keynesian theory was that internal public debt has no limits if it is required to fill up demand deficiency. Borrowing could always be resorted to pay back old loans without affecting the monetary equilibrium. Through open market operations a central bank could keep both money supply and inflation order control. This is precisely what Aba P. Larner and, Alvin Hansen, the two well known names of the time argued. Of course, not even the most dogmatic Keynesian is prepared to touch that idea now but enough damage had been done to developing countries.

The idea of a self-perpetuating internal debt excited the economists from the LDCs because it seemed to provide a painless method of financing the public sector is which they had put a profound faith. After four decades of increasing reliance on fiscal deficits and public debt, we find ourselves caught in a debt trap or a fiscal vicious circle from which it seems difficult to get out without drastically reducing both Government expenditure and public borrowing. A high debt-GDP ratio, let alone a rising ratio, acts as a push towards disastrous monetary and economic imbalances.

Four facts pinpoint danger signals. First, the total debt of the Central and State Governments now stands at about seventy per cent of the GDP. Second, internal and external debt are running neck to neck, one reflecting the internal and the other external imbalance. There are many points of interaction and mutual re-enforcement but every linkage is not fully exposed. For instance, it is not often realised that external loans require a corresponding creation of rupee funds for domestic absorption. What is put in the external funds is used to pay for imports. Thus, the larger the external debt there is bound to be a larger requirement of internal borrowing. In recent years, the direct tax rates have been slashed and the rates of tax collection to GDP have declined requiring internal borrowings. But the main point is that foreign aid acts as an independent variable and promotes growth of domestic debt.

Third, the most negative development of the eighties, as against the earlier decades in which public debt was incurred to finance capital expenditure, it was the revenue deficit whcih came to be financed partly by public borrowing and partly by created money. It has been lost on the Government and the economists that in the short run both can have nearly an equal impact on monetary expansion by the banking system. The Reserve Bank was unable to stop the monetization of the debt. Fourth, the argument that the Government creates assets against borrowing was not fully valid in view of the above mentioned fact that borrowing was partly required for meeting deficits. Equally if assets created did not earn at least the interest paid on the debt, then there was an inbuilt acceleration of internal debt. India presents to the world a model of unmitigated anarchy.

Chapter 5

Balance of Payments: Dependency and Deficits

Balance of payments and foreign exchange crises have intermittantly plagued the Indian economy ever since 1957. But never before did the crises come crashing on the Government and the economy without a solution. Short term crash borrowing from the IMF and other agencies instead of improving is further contributing to a deterioration of the situation because no domestic adjustments were undertaken until recently. The threat of a slowdown hangs over the economy. India's reserves are down to perilously low level. India's credit rating has fallen to junk bond status. International institutions and the larger developed countries have put India on notice.

It was the desperate act of a desperate Government to devalue the rupee under circumstances that did not warrant it. Neither was the global situation favourable nor were the import-export relations such to justify devaluation. All the arguments advanced in its favour were standard IMF conditionalities intended simply to get a large loan. The question is whether without removing the causes that led to payment deficits, a large debt will not lead to the repetition of the same problem.

For over seven five-year Plans, the external components of fiscal policy were wrongly assumed to be rather a passive addition to our resources and budgets. There were strings attached to every penny of external finance, yet our leaders told us it was not so. We could have partly minimised these intrusions only if the ministers of External Affairs, Finance and Commerce acted as a team as did Korea and Japan.

Though the external crisis is deepening and terrifying, the drum-beaters of the now failed Nehru-Mahalonobis model and the guilty men of India who still control strategic institutions of thought are adding insult to injury by suggesting that the external crises has been due to exogenous shocks. They refuse to admit flaws in fiscal and industrial policies, import-export strategies and absence of self-reliant technology policies. It is no accident that five-year Plans had the blessings of nearly a hundred foreign economists of both left and right pursuasions who joined their local clients in a conspiracy to defraud India of her technological independence, particularly through a fast expanding licence-permit raj and in the name of state correcting the market imperfections, and who are now shouting hoarse for liberalisation.

The high growth rate of the eighties was particularly achieved at the cost of heavy dependence on external savings. The external savings ratio changed from 1.5 to 3 per cent of the GDP. These savings which took the form of import of goods were acquired on conditions that gradually impinged on shifts in domestic policies, particularly in respect of capital goods, consumer durables and defence. The impact was far more serious on the pattern of industrialisation, import intensities of industries and pattern of energy consumption. All these impacts reduced the viability of our external payments.

The ratio of current account deficit to GDP averaged 2.2 per cent during 1985-90, far above the 1.6 per cent projected in the Seventh Plan document. But in the last year of the Plan and 1990-91 it increased to 3 per cent. On the other hand, the volume of export which increased by 10 per cent per annum during 1986-87 to 1989-90 also came down subsequently. The real problem is that as much as 50 per cent of the nation's financing needs are met by external aid from international institutions and non-resident Indians and market borrowing. Therefore, any policy which is of short-term nature and aims at increasing external aid can be self-defeating in the long run. The BOP crises and the sharp reduction in foreign exchange reserves should be turned into an opportunity to completely reorient the industrial, fiscal and commercial policies. All the three have to be packaged. Since the public sector is the largest consumer of imports, the losing concerns should be drastically dealt with. Most significantly, India should start picking up new trading partners for collective action. If all this is not done, efforts to get extra external resources will along with devaluation of the rupee produce a new crisis.

My colleague in the Planning Commission Arun Ghosh noted "for the most part, the impact of the import of capital goods has, of late, been of two categories, items which are imported merely because they are available under external aid programmes (and our internal resources are not enough to finance domestic procurement even though we may be competitive in the production of small items); and recently, capital goods, which are required to be imported because of the incorrect focus of our industrial development programme" (EPW Feb., 9, 1991). In other words, our payments gap is, both explicitly and structurally, the product of two vested interests: the aid seekers and industrial policy makers.

A Long Drawn Vicious Circle

Indeed, the persistent balance of payments crisis is an expression of yet another long drawn vicious circle. In this case, let us begin from the end rather than from the beginning. There is an ongoing controversy as to whether or not and how far to externally liberalise the economy in order to make Indian goods competitive in the global market. Under a variety of internal and external controls, the high priority to import substitution and low priority to export promotion led to a decline in India's share in the global trade in four decades from 2.5 to 0.4 per cent.

Even now, exports stay at this level despite many pushes and subsidies given to it. For a long time the balance of payments deficit was interpreted not as structural problem but the counterpart of foreign aid and it was assumed that once the economy became self-reliant, there would be no need for aid and thus the balance of trade would arrive at its natural balance. The logic of this analysis depended upon certain assumptions, none of which was fulfilled.

First, if the share of Indian trade was to decline to such low limits as it did, obviously it must have been implicitly assumed that there would be a certain delinking of the Indian economy from the global system. Second, the same logic also implied that the external component in Indian development would decline proportionately even if it increased in absolute figures. What happened was quite the contrary. The import intensity of Indian industries in terms of both capital and materials increased. However, the rate of domestic production of these imports was not upto the desired levels, precisely because of the producers' preference. There was no policy of creating equilibrium between the import content of domestic development and the development content of the import. Third, it was falsely assumed that import substitution and export promotion were always mutually exclusive. Unlike Japan and Korea, we failed to comprehend that over a wide range both were mutually reinforcing; it was only a question of right sequencing. Fourth, whatever the strategy, controls and regulation along with the fiscal system turned Indian industry into a high cost economy. Instead of breaking monopolies, controls provided producers with a sheltered market and a corrupt system which made monopolies come through the backdoor of restricted entry.

One by one, these assumptions collapsed. We can now return to the first principle. The understanding that the balance of payment gap was simply a reflection of foreign aid, though somewhat correct, led to a mounting foreign debt, even though India got a large share of aid at concessional terms. Aid was not optimally used and could not be used so in a highly corrupt, restrictive and restraining system. So much so that not only the servicing of debt rose to as much as thirty per cent of exports, if we include the fact some part of new aid was utilised to repay old loans. As a result, the balance of payments deficit widened as did the need for more aid and imports. Consequently, the foreign exchange reserves shrunk to a few weeks' imports. Resort to IMF became inevitable. Indeed, repeated resort to IMF has now become a necessity and will remain so long as the structure and pattern of industry do not change and the regime of low productivity does not come to an end. External debt is bound to go up as we take more loans to pay for old loans. A new vicious circle has been created.

Dr Arun Ghosh summed the position thus: "the industrial investment programme in the Seventh Plan had little or not impact on our exports. If exports suddenly improved, they were the cumulative result of other policies like the steady depreciation benefits accruing to exports". In other words, whereas industry became more import-dependent, it has little export-orientation despite all the export subsidies and policies to stimulate it. The balance of payments widened, particularly in view of the technological dependence.

Once the government opted for liberalisation in 1981, the demand for liberalisation became both vocal and irresistable. The Rajiv Government allowed external liberalisation, instead of first going for domestic liberalisation. The structural imbalances and vicious circles continued to operate. As a result, the more the liberalisation, the larger became the deficit and still more strident the demand for liberalisation. The rupee kept on depreciating and though it encouraged exports to some extent, it ultimately pushed up the rupee cost of foreign debt. India is already in a foreign debt trap or vicious circle of "imported" import substitution, export lag, dependence on foreign aid, declining share of trade, drying up of exchange reserves, demand for liberalisation and rupee devaluation. We are back to the balance of payments crisis. The vicious circle is complete.

A continuous devaluation of the rupee—both implicits and explicit—was thought to be a less painful method of adjustment than straightforward one-shot large devaluation. The rupee has been depreciating at the rate of 10 to 11 per cent per annum. But it seems the path chosen paved the way for the balance of payments crisis. There is no clear relation between the nominal and real exchange rate. The continuous devaluation did not change production levels and the ratio between trade the non-traded goods hence did not remove domestic constraints in production.

Balance of Payments: Dependency and Deficits

Most significantly, as was feared, a continuous devaluation created expectations of future depreciation which indeed acted as a powerful pull for increased dependence an imports and inventory built-up, all in the name of exports. It was forgotten that the implicit devaluation would not stop the explicit devaluation under international pressures, if it did not improve the trade position.

Probably the most important reason for the widening balance of payments deficit is on Government account and not on private account. The larger the investment in the public sector, the greater are the imports as the import content of big projects often financed through loans is very high. The performance of the Government sector on exports remains dismal.

The Most Potent Factor

The failure of domestic demand management has been the most potent factor in creating the balance of payments crisis. The Government has been overtrading the commercial policies, often wrong and delinked from domestic demand, to solve the external problems. Faith in the continuous and implicit devaluation of the rupee to resolve the contradictions proved unfounded. The dangerous potential of relations between the current account and exchange rate was ignored. Elasticity optimism was unwarranted.

It is amazing that a country boasting about planning had next to nil incomes and expenditure policies. There has been no check either on the Government consumption expenditure or on the consumption of the non-poor. The poor people were squeezed by inflation arising from excess demand of other classes, decline in employment growth and, even more significantly, from the wealth effect of both capital-intensive investment and inflation. Both the public and private sectors freely indulged in conspicuous investment and two oil shocks did not prompt them to rationalise energy consumption, with the result that the oil bill continually went up. The policymakers paid no attention to the costly transfer effect of rise in oil prices. Instead of adjustment, the government looked for avenues for external financing. The dependence on external financing increased further as the terms of trade became adverse requiring further dependence on external financing.

Every member of the Planning Commission knows that he is an unwitting tool of the Government for expanding its consumption as much as capital expenditure. When a new five-year plan is prepared, increase in consumption expenditure at the rate of 10 to 20 per cent is explicitly provided. The Commission never stipulates external financing of such expenditure. What happens actually is that the Government always looks for maximising external resources in order not to raise them domestically and asks the Planning Commission to accept hundreds of proposals of expenditure for which specific foreign aid is available and provide rupee funds from the Plan funds. In hundreds of proposals and projects, some of which are not included originally in the Plan, aid becomes the motive force for expanding Government's corruption expenditure. Most of the money is pocketed by corrupt bureaucrats and politicians in respect of socalled soft sectors. But, the external burden keeps rising for which there is no provision to repay except by borrowing more of such aid. The Planning Commission is used by the Government as a fig leaf for legitimising its sin of pushing the Government budget to external dependency. Internal and external policies reinforce one another to create a structure of external dependency. People don't realise that the populism and an anti-poverty programme lacking in commitment to remove of poverty were the major cause of the external crisis as it was partly externally produced.

Indeed, there has been a double menagement failure, internal as well as external. The Government borrowings from the Reserve Bank for the budget rose from 20 to 30 per cent in the last decade. The persistently high fiscal deficit resulted in the total liability of the Government rising from 44 to 60 per cent of the GDP during the eighties. Money supply grew at an average rate of 18 to 20 per cent creating serious inflationary pressures. The current rate of inflation is nearly double that of a few years ago. The current account deficit in external trade and payments reached nearly $9 billion, about 3.2 per cent of GDP as against 1.2 per cent in the earlier decade. The external debt outstanding has crossed $70 billion. This figure does not include some private debts. The debt service ratio, which remained at 25 per cent

Balance of Payments: Dependency and Deficits

despite growth in exports and a reduction in repayments of IMF credit, is slowly increasing, as old loans got at a concessional rate are replaced by those with high interest rates.

The worst part of this vicious circle is hidden. In the discussion about the balance-of-payments, there is no mention of the capital flight which has now assumed enormous proportions. Balance-of-payments statistics officially recorded have no flows between India and the rest of the world. Some were not officially recorded, others mis-recorded, still others were concealed. Above all, the miscalculation of invisibles cannot be easily interpreted. The former Finance Minister, Yashwant Sinha revealed that there were some confidential accounts which he had to tap to meet the fiscal crisis. He did not give any figures but according to knowledgeable sources, the outstandings were as high as Rs 5,000 crores.

There are many ways in which money is illegally transferred from the developing countries to the developed world and by which the operators on both sides enter into a variety of dubious deals to defraud the poor nations. Often one section or the other of the Government and the financial institutions is involved in this game. It is not possible to underestimate the impact of this phenomenon of illegal transfers and transactions. They are now overtaking financial policies completely. For instance, it is said of Brazil that its foreign debt of about $100 billion is fully matched by illegal deposits of the Brazilians lying in the banks of Miami, New York and Switzerland. The consequences of these illegal activities are for all to see. The Brazilian economy is caught in a debt trap, creates inequalities, is inflicted by hyper inflation, which is hundred times faster than the growth of the economy. It is a democracy which remains constantly under threat and above all, has fallen into an almost permanent dependency situation.

Brave Statements

India is not far behind Brazil now although we keep hearing brave statements from Government leaders, planners and economists, that our policies are designed to stop India from falling into a Brazilian trap situation and for giving ourselves self-reliance. Now that the financial crisis has become deep, serious and irrevesable, we are beginning to realise that we are in no better position than Brazil. In fact, Brazil's per capita income is four times that of India and gives her a cushion for absorbing the crisis. Brazilian and Indian poverty situations are not comparable.

Statistical figures never tally. On our side, there is a wide discrepancy between the trade accounts prepared by the Reserve Bank and the Directorate General of Commercial Intelligence and Statistics (DGCIS). However, there is counter checking of these figures by the trade statistics of the importing and exporting countries. We have also some information from the international financial institutions and economists have worked out methods by means of which they can provide rough estimates of the flight of capital from one country to another. It is well recognised that the estimates for external indebtedness have often had a large element of underestimation of capital flight. However, these figures are revised subsequently.

According to Rishi and Boyd, between 1971 and 1986 the capital flight from India taken cumulatively came to about $28 to 29 billion when the external debt was of the order of $40 billion. There is no reason to believe that any decline has taken place since then. Indeed, if one goes by the difference between the official and non-official exchange rates of the rupee, capital flight must have increased further. In 1990, the external debt of India was $70 billion. In other words, somewhere between $40 to 50 billion of capital flight should have cumulatively taken place.

When the crises exploded during and after the recent Gulf war, the Government panicked. Domestic political uncertainty made the creditors withdraw their hand of help. In sheer desperation, the Government went on a borrowing spree but the response was jerky and dismissive. The current fury of crises and loss of confidence about the expected future bounced back to create an even greater panic. Even gold reserves were mortgaged.

In the wake of the balance of payments crisis came the costly manipulation of import licences, replenishment licences and exchange transactions which not only distorted trade but also put pressure on the rupee. Above all, it made nonsense of the planning of import-export equilibrium. There are not more than a dozen cartels of brokers and speculators who have taken over the import licence market and fleece the genuine importers who are obliged to ask for new licences of larger values, compounding the already difficult situation.

In the past few years the World Bank and our business community strongly advocated substantially higher commercial borrowings from the capital market from the present level of $1 billion to 2.5 billion. The borrowing limits were expected to be raised upto $5 billion but the project had been overtaken by the crisis itself. However, the arguments given in its favour were the following:

(1) Access to traditional sources of soft loans like that of the IDA is progressively declining; (2) There is a vast European market for capital of which India has not touched the fringe even though the market is expanding and has much surplus funds; (3) The smaller countries which have been success stories borrowed more than India did. Examples were cited of Korea, Malaysia and Indonesia. All these were attempts to postpone the evil day, i.e., shift the dependence from the official to the market sources.

India's credit rating slumped to the lowest point. It looks as if there is a conspiracy between international and national credit rating systems. The international system has to protect itself against the misuse of their money given to Third World countries but when the Governments of the latter use the same method to do internal rating, they fall into the trap of the international finance because of the advantages the foreign owned and foreign collaborated companies enjoy and which automatically give them high credit rating. Three years ago, the Credit Rating and Information Services of India Limited (CRISIL) promoted by ICICI introduced the concept of credit risk evaluation for corporate debt investment as developed in US and Europe by such agencies as Moody. We are not concerned here about the techniques of evaluation. Normally, one would welcome such an agency but one look at the tables presented by CRISIL would show that this ranking has become a conspiracy. The companies which have been given AAA are (a) big business houses compnaies, (b) FERA companies, (c) Indian companies which have large collaboration in terms of finance.

The desperate attempt to plug the balance-of-payment deficit has now reached the absurd situation of deliberately mortgaging our future. The Government has persuaded the international commercial banks to make advance payment against future exports of bulk commodities, such as iron ore, coffee, leather, gems and jewellery. The Australian-based Grindlays Bank has been entrusted with this job. Such an arrangement means that the future free earnings will be less.

The shift from colonial relations to independently determined external relations was reduced to a few rules and regulations and a new kind of dependence on foreign aid without taking precautions on many kinds of imports which relinked the Indian economy to old colonised relations. The central problem is one of sound-economic transformation and institutional change and not of a few simple shifts of policies. Goods had to be imported and exported but both with the structure of international trading regimes and consistent with the objectives of national development. The crisis is historical and structural.

The balance of payments problem cannot be confined to adjustment in the field of macro policies. The deficit is an instrument for throwing the burden of adjustment on the poor. The distributional question is seldom raised. For instance, as pointed out by Pulapre Balakrishna, "the poor in India are now being asked to pay for essential imports (or to even go without them for a while) so that the national economy can repay debt incurred in the purchase of aircraft, among other things". (E.P.W., Feb. 16, 1991).

CHAPTER 6

A Vicious Circle of Export and Import Substitution

Over the last 40 years, two strategies, totally delinked and one coming in the wake of the other after a very large time gap, have dominated Indian economic development: import substitution and export promotion. Both assumed certain but unspecified links of the Indian economy with the rest of the world. It is an important assumption of all strategies of development to earn foreign exchange to pay for the import content of development. On the face of it, it also seemed unexceptionable that goods previously imported during the imperialist phase should be produced at home. The question that was ignored with a frightfully heavy cost paid was whether all the goods previously imported, including luxury goods, had to be substituted by domestic production. It is no accident that we have now arrived at a situation in which the growth of the consumer durables is much faster than that of any other category of manufactures. Ironically, the single most important factor to which the very first crisis in the late fifties of the Indian economy was traced was the delinking of import substitution from both export promotion and autonomous development of the economy, so much so that ultimately import substitution became the chief determinant of development.

Similarly, it is equally unexceptionable that we should maximise our earnings by export promotion, though we did not have to subscribe to the slogan of export or perish. But now, by ignoring another crucial question, we may be making the same mistake again. Do we have to allow the general pattern of our production to be determined by the demand generated from external sources and supplies constrained by imported technologies? After all, the export obligations imposed on industries form only a small part of their production and relate to a fixed period of time after which their products get absorbed in the domestic demand structure. If export promotion is delinked either from import substitution or from the main development strategy, the consequences will be precisely the same as were in the case of import substitution.

It is worth looking into the question why import substitution did not achieve the aforementioned results and exports did not go up when the import substitution processes reached an advanced stage and even got exhausted and produced distortions, stagnation and crisis. This happened largely because of the contradictions in the process of import substitution itself.

Import substitution also created a structure of incomes which was highly skewed in favour of higher deciles. It relied on a high capital-output ratio and a capital intensive-technology which were bound to create concentration of incomes. Whether new consumer goods industries were large or medium, the incomes created and goods produced were largely for the upper and middle income groups. For a while, possibilities for profit maximization were opened up in the field of consumer durables. The size of the demand remained adequate for a while. But because the incomes continued to get concentrated, the demand too became concentrated in narrow, upper and middle classes.

Most important, the production of these goods was allowed a fully sheltered market arranged through a complicated set of controls and regulations. This sheltered market provided no incentive for technological changes except those which came through foreign collaboration under highly constrained conditions. There was no pressure for building either domestic or external competitiveness. The export market did not and could not expand sufficiently when it went along with this pattern of the domestic production and demand and hence industrial expansion slowed down after a decade. Under the shelter of high tariff walls, high

A Vicious Circle of Export and Import Substitution   49

profits, high prices and unutilised capacities continued to exist side by side. Now, the same may result from export promotion if it remains divorced from other relevant strategies.

Import substitution cannot be considered in isolation from the structure of tariffs. Indeed, given a general positive policy of import substitution, differentiation in tariffs determines the structure of import substitution and hence of the industry itself. The Government, the industry and foreign collaborators formed a network together to share almost the total benefit.

Maximum Protection

Indeed, by giving maximum protection to and imposing the highest tariffs on import of capital goods and raw materials only after the natural spurt of imports, the Government was obliged to give almost total protection to consumer goods industries. On the face of it this policy had a strong bias for self-reliance. Import duties on capital goods and raw materials were either non-existent and when levied for revenue purposes, were less than one-third of the value, in contrast, the duties on the import of consumer goods were several fold of that on capital goods. The latter was also subject to quotas. The reason for this discrimination was to reduce the cost of laying the industrial base. There was ample justification for this policy but by ignoring the consequences of such a policy and not providing countervailing policies, the structure of industry and the relative position of consumers and capital goods industries were distorted. Instead of promoting capital goods industries it promoted durable consumer goods industries.

The pricing policy of industrial products also proved innimical to the growth of capital goods industries. There has been very stringent and often economically unjustifiablke controls on the prices of such basic industrial products as steel, cement, aluminium, coal, chemicals, fertilizers and bulk drugs, leading to their uneconomic prices as well as blackmarketing. On the other hand, there was practically no price control on consumer durables or when controlled, the result was black market and black income.

These pricing policies, along with other wasteful practices in the public sector projects undermined the very structure of the public sector, leading to its mounting losses, unaccountability, mismanagement and gross inefficiency. The decline in the profitability of public sector proved to be the most damaging factor that retarded the growth of capital goods industries.

Import substitution also lost the capacity to lower the average import coefficient, taking into account machinery, raw materials and energy consumption. That is why the growth of imports is still constrained by aid and trade or by the capacity to import by earnings from exports. Exports too depended on a large import component, without, in effect, raising exports significantly. The recent balance of payments crises have revealed both these structural deficiencies. Consequently, India has faced a double disadvantage: the stagnant industrial growth and increase in foreign indebtedness, both on private and public accounts.

Since import substitution had to be given the highest priority in a vast field of consumer goods the policies had to be differentially conceived. The consumer goods can be divided into four groups: (1) those strictly desired for exports, (2) those required for domestic mass consumption, (3) consumer durables, and (4) those which were being imported historically. Import substitution had no significance for (2). For (3) where it was most relevant; a strict order of priorities and constraints was required. Unfortunately, (3) and (4) were given the highest priority as it satisfied the demand of middle and upper classes in their country. Somehow, this wrong policy was accepted and rationalised into a general strategy for accelerating industrial investment and growth.

A more sophisticated assumption underlying this approach was the acceptance of some kind of a quasi-Keynesian approach according to which investment, without qualification, was to create its own savings. This assumption is no longer valid. In fact, it was never fully valid as it is known now that the rate of savings in the industrial corporate sector had always been low and the domestic private savings in general, after having reached their highest point in mid-sixties, stagnated and even declined afterwards never to reach again their earlier higher level. Once savings reached a plateau, the main function of import substitution of encouraging investments also lost its force.

A Vicious Circle of Export and Import Substitution

Import substitution has to be distinguished in its two eventually very different phases, as the early phase and a later phase. Most significantly in the latter phase, the sequence has had to change with export promotion getting a higher priority, though their linkages were never be disrupted. In India, the problem has been both of wrong linkages and complete misperception of the sequence.

It is expected that in the earlier phase, import substitution generally leads to (1) saving foreign exchange, (2) pushing up of investment and savings partly by putting restrictions on imported supplies, thus leaving demand unsatisfied, and partly by directly encouraging investment in the domestic production of those commodities, (3) encouraging foreign companies to invest in developing countries to enlarge their share in the latter's market, (4) raising profits in new industries. However, this process has to be reinforced as well as corrected by exports at some stage as Japan and Korea did and we did not. The sequence was broken and the Indian industry-export-import ran into a vicious circle, of larger the imports, the smaller the proportion of exports, the larger the production of consumer goods the higher import intensity, greater the imports the more stagnant industrial growth itself. The vicious circle was very vicious indeed.

"Imported" Import Substitution

The Indian import substitution, in contrast to that of Japan and South Korea, essentially remained an "imported" import substitution and ultimately became a kind of export substitution. Imported import substitution means dependence on imported technologies almost totally and repeatedly, instead of accompanying imports by its indigenisation, adaptation, modernisation and after a while, its complete replacement through domestic R and D. Importers of technologies were not obliged to show results in indegenisation. Surprisingly, they were freely permitted to import and from specified sources as a result of the way foreign aid was tied and administered. As India was flushed with aid acquired at cheap rates in the early decades, she let the importers import whatever technology they got and of whatever quality but under conditions that thwarted development of domestic technologies.

The growth or expansion of capital goods sector for which technology imports should have been given the highest priority lagged behind. What kind of technology could an affluent society offer? It was most in consumer durables or capital goods for producing capital goods domestically. If the growth of the latter slowed down, it was because of import policy itself. The infrastructural technology was not always neutral between the two but given well structured policies it could be neutralised. Airlines could wait until railways had achieved the set targets. However, the same mistake is being made in export promotion. The emphasis is on the export of consumer durables in which late-comers and technologically capital dependent countries cannot compete. Hence, the crises of macro imbalances.

The greatest and most dangerous illusion created by the business and Government acting naively or conspiratorially and supported by the drum beaters of the Nehru-Mahalonobis model was that import-substitution was autonomous and self-reliant. In reality, import substitution was, instead of being trade-promoting, aid-induced. Thus, it slowly and persistently built up external obligations. Aid-induced import substitution seemed an easy option as it imposed no obligation to pay for imports needed for production of exportables. The much needed differential pattern of investment and constraints on the constraints on the consumption of the non-poor were substituted by the indiscriminately liberal imports policy. The rulers preferred to pursue the *status quo* and let their class have the best of both worlds, high savings as well as higher consumption. Since aid in the early decades was at a very low rate of interest, the rulers showed a higher degree of irresponsibility by discounting the future obligations. When the chicken of repayment came to roost, the whole edifice crumbled. Had the same growth rate, which was quite modest, was engineered through exports rather than aid, as the Japanese and Koreans did, there would have been no crisis, no need for unnecessary import substitution, and, above all, a global niche made available for our exports in this highly competitive world. It is too late now for exports to jump. Only a slow growth is feasible.

Nations such as India which did not pursue export-orientation

A Vicious Circle of Export and Import Substitution 53

as part of a natural evolution of the development pattern and appropriate policies thereof were condemned to adopt "vigorous" export-promotion later on. Had they thought of import-substitution and export-promotion as two sides of the same problem right at the beginning, subsequent crises of growth of foreign exchange could have been avoided. What the Government tried through liberalisation was once again the same kind of dichotomised approach but in reverse order. The Government aimed at export promotion by inviting large foreign investment at the cost of import substitution. Without strengthening domestic competition sector-wise, liberalisation of imports hurt import substitution. Both phases thus deepened the crises. The import of technology in the first phase was focussed on infrastructure and consumer durables.

There are several reasons for which the export prospects seem to have been suddenly dimmed after last year's exceptional increase. First, the general expansion of world trade to which the trade of every country is linked is expected to slow down either because of continuing recession or because of the aftermath of inflation of the last three years. Some recent estimates for the US economy suggest that the recession may not be on its way out but it is too early to make definite projections. Anyway, Indian exports are unlikely to get any major boost in that market. Second, the policy of stockpiling which many countries adopted following the 1973 oil crisis has run its full course. This will seriously affect the exports of weaker economies.

Third, the developed countries have made necessary technological breakthroughs to find substitutes for the goods earlier imported from the poor nations, thus reducing their dependence on the latter. Fourth, the organisational, managerial and technological changes in the advanced countries obliged them to trade and invest within the closed curcuit of their own kind of nations. The poor countreis were put on notice that their export possibilities must decline. Finally, the collapse of the Eastern European communism and rush for market economies have clearly given another notice to the poor nations that they no longer were on the top of the list of priorities of the developed nations. More resources will have to go to Eastern Europe.

After the payments crisis that burst out during and after the Gulf war, the Government and the Reserve Bank drastically slashed imports as well as put the most stringent restrictions on bank credit to industry. The result was a sharp reduction in industrial production and damage to infrastructure which now indicates a lower future growth rate.

Two outstanding features of Government's current economic policy are (1) a credit squeeze designed to keep price stability with all related objectives pushed to the background for the time being and (2) a massive liberalisation of export incentives. Everything else seems to be subordinated to them. The two are independent policies but they are also related in many ways. There can be no such thing as a long-term credit squeeze. In the short run, the two policies can either reinforce or cross-cancel each other. Since credit squeeze generally does not apply to export promotion elsewhere, the squeeze has to be carried to practically those prohibitive limits where it must inevitably affect domestic demand if the relationship between money supply and incomes is to be such as to stablise prices. If the rate of domestic production is adversely affected or if it does not increase but, on the other hand, exports are pushed up, the credit squeeze may produce results opposite to those intended. If this happens the export strategy too will have to be revised.

If export earnings remain stagnant or increase slowly but the demand for capital and raw materials, particularly oil, goes up, then any further reduction in imports which are of high priority and are not wasteful will surely lead to an intensification of the internal stagnation and instead of increasing will reduce the utilisation of existing capacity for export promotion. Thus, it seems that there is not one but numerous vicious circles in the current situation.

A Very Big Constraint

A very big constraint on our export promotion is the very small surplus of exportables within the country. Consequently, exports

A Vicious Circle of Export and Import Substitution

cannot be increased beyond a point without reducing domestic supplies to those critical points at which prices are bound to shoot up. On the other hand, if prices are artificially kept down through intensification of the credit squeeze, the consequences can be even more serious for future production. Paradoxically, that magnitude of credit squeeze which can really keep prices under control for a reasonable length of time can do so only by reducing production. Why does such a correlation exist? The answer lies in a rather more fundamental economic restructuring.

Despite everything that has been done by way of export promotion and changing the pattern of industrial production, it still remains a fact that only a few important commodities dominate our exports. These are sugar, engineering goods, jute and jute manufactures, cotton textiles, gem and jewellery, leather and leather goods accounting for more than three-fourths of the total exports. It is precisely in respect of these commodities that export prospects are not very encouraging.

The real difficulty is that there is still not enough co-ordination between that part of the industrial policy which is directed towards maintaining a particular industrial structure at home and that part of the structure which is designed to promote exports.

The export strategy which was evolved in an entirely different context, namely of price stability and domestic available capacities on the one hand and a rather inelastic demand for Indian commodities abroad on the other, is no longer relevant to the new situation except in the case of the last mentioned problem which is not in Indian control anyway. In its attempts to reinforce the export strategy and to make it more responsive to fast changing international developments, particularly the rise in the prices of raw materials and machinery, the Government is, perhaps unconsciously, aggravating the problems of domestic inflation.

A more important distinction is between items the export of which is largely non-inflationary and those for which the creation of an export surplus will inject a high degree of inflation if the market mechanism is allowed to work. On the basis of this latter distinction, commodities for which there is a world demand and which can also fetch good prices are those for which, unless the entire supplies are within the control of the Government, exports if artificially pushed up add fuel to the fire of domestic inflation. Already such goods as leather, oil cakes, cotton cloth, fish, coffee, spices and vegetable oil, have pushed up domestic prices, intermittantly, we also export rice, sugar, cement, paper commodities for which there is a good market but all of which are either in short or uncertain supplies at home.

For keeping control on the supplies of essential commodities for domestic consumption as well as for exports, the Government largely relies on dual pricing policy. This system can function only if the autonomy of the two prices—the controlled price and the open market price—is maintained. In a situation of prolonged inflation and increasing shortages, this autonomy cannot be maintained because the open market price dominates and becomes the real price. Even a minor diversion of supplies from domestic to export markets and inject a large dose of inflation.

To meet the balance of payment crisis, the government has evolved a four-fold strategy: (1) get as much long-term and short-term aid as is available from whatever quarters, particularly the import of crude oil on credit; (2) offer a variety of incentives to exporters at considerable cost to the exchequer; (3) increase exports which have a demand abroad but which also require curbs on consumption at home; and (4) secure higher unit value for the main export items by raising export prices. The danger then is that either the new export strategy may fail or if it partially succeeds it will deepen the domestic inflationary crisis. Not only is there no co-ordination between various domestic and external policies, there is also no coherent national policy to generate export surpluses in such a way as to avoid the widening external gap and the inflationary impact on the economy.

A large number of measures have been taken to liberalise the economy but these measures do not add up to a coherent and well recognised policy. Some call it adhocism, others call it step-by-step strategy. Whatever be the case, a number of steps are still to be taken before a clearer picture would emerge. The belief that for three decades, the broad policy adopted was import substitution as against export-led growth i.e., growth through import substitution and not through promotion of exports is half-truth. This is because whatever the policies, these were ineffectively pursued. The Abid Hussain Report has talked of growth-led export instead of export-led growth as the model for India to follow. Every economist knows that there can be a policy or a model of export-led growth but to say that it is a growth-led export will not be correct because growth stands for a total sector growth and not a particular sector's growth.

Despite these and other concessions, the gap between imports and exports has widened. Again, leaving out all serious fluctuations the excess of imports over exports which was about 10 to 12 per cent of the exports in the mid seventies has now risen to about 40 to 50 per cent of the exports. The Government has to take into account the consequences of its policy of liberalisation on trade and the economy. The way liberalisation is being implemented is not likely to create a larger exportable surplus in any substantial way. Nor will it be able to force the big producers to export. It is because international competitiveness based on comparative advantage is preceded by other autonomous policies. For instance, South East Asian countries "maintained their protectionist import substitution policy package by moving into capital, durable goods and processing areas; their industrial exports have not been strongly dictated by considerations of a dynamic comparative advantage" No wonder, the impact of our policy on employment and distribution has been negative. The decision centres of the Indian economy have moved outside India, be it World Bank, IMF or any other organisation. India is entering a new phase of dependency, slower growth and debt trap. The siege is still on.

The drum-beaters of the Nehru-Mahalonobis model could cover the miscalculation of the import dependence of their plan by taking refuge under foreign policy on which fell the entire burden of procuring aid. We cannot go into the distortions India's foreign policy has been responsible for, but the tragic part of the story is the marriage between foreign policy and the Mahalonobis model conceals the gruesome selfishness or lack of understanding on the part of the new ruling class.

EC 1992

Finally, prospects for trade after 1992 are even more dismal. The ADC's, including both US and Europe, are demanding total reciprocity from the LDC's, thereby the entire basis of competition on the basis of limited rights for protection enjoyed by the LDCs. Competition is fine but with the ADCs' total control over technologies and future technological developments, competition will always remain unfair. It is doubtful if even those sectors of the economy which managed to sustain with imported technologies will survive. EC in 1992 will be the single largest trading entity which in competition with USA and Japan will resort to all kinds of protectionist measures that will hurt the Indian and other LDC economies. A document of the German Development Institute says "There is a risk that the common trade policy of the EC could become even more protectionist after 1992 than before. A number of European governments seem to believe that they can only manage the risks of this single European market behind effective protectionist barriers against the outside world".

It says further that "Third world countries, most affected by the single European market will be the newly industrialising countries. These countries, mainly which will be China and India, have already geared their export structures accomodate more manufacture than all the other developing countries. Their industrial exports might suffer from increased competitiveness of Europe's industry and the trade diversion effects of the elimination of borders—the more so, since East European countries will be exporting more low cost manufactures to West Europe". Finally, "With their limited adjustment capacity to changing world market conditions countries like India and Egypt will encounter serious problems in keeping up with the stiffer competition in the European market".

Other aspects of the global trade which have over the years undermined the position of the LDCs are (a) the growth rate of volume of world trade in agricultural products declined from 4.0 in 1960 to 1.5 per cent during the eighties, (b) the trade in minerals declined sharply from 7.5 to 1.5 per cent, (c) the growth rate of trade in manufactures doubled from 4.5 to 10.5 per cent

A Vicious Circle of Export and Import Substitution 59

in the same period, (d) world trade in agriculture raw materials such as cotton, sisal, jute and wool has declined in absolute terms by 5 to 10 per cent in the last twentyfive years.

The reasons of these trading trends which have adversely affected the LDCs are many. For instance, the substitution of synthetics because of technological developments for dematerialisation (products becoming lighter and small), decline in the mineral-intensive smokestack industries, and ecological considerations are some of those listed. But there are other more pregnant reasons which reveal terrible prospects for the LDCs. The ADCs mainly trade among themselves and invest within the ADCs (completely falsifying the Marxist theory that finance capital must look for investment outlets in LDCs. These lead to a reverse transfer of resources from the LDCs to ADCs, a declining share of LDCs in global manufacture and above all, the widening technological gap between the two.

The theories of imperialism, interdependence and dependency are all being proved obsolete one by one. They are being replaced by the theory of International Economic Darwinism. Colonies and peripheries have served their purpose. The Marxists, the Brandites, the Preleishites, the Nehruites are all unable to explain the phenomenon in terms of their theories.

The trade theorists who had been pushing for trade order for years and who were in the service of international institutions as puppetteers, manipulating the mandarins and economists of the LDCs have come to admit that trade equilibrium, in the conditions of trading with multinationals, uncertain technologies and trade concentration, can be achieved only with and not without tariffs. The unresolved question is how much of tariffs. Unfortunately, Jagdish Bhagwati's puppetteer governments and his tribe can not produce such equilibrating tariffs.

Chapter 7

A Decripit Public Sector

The Government sector, which now generates about thirty per cent of the GDP, incurs an expenditure of nearly forty per cent of the GDP. This is a major source of crisis as this expenditure is increasingly financed by borrowing and created money as against taxes. There are two other activities that are often confused to be one. The first is the issues and size of the public sector business enterprises in ratio to those in the private sector and the second is the Government controls and regulations versus the role of the market. Both have been the subject of debate without changing the terms of the debate and that is why there are not settled points of consensus. The protagonists of the Nehru-Mahalonobis model who set the terms of the debate are still arguing from their bunkers, notwithstanding the fact that the Soviet Union and other Eastern European countries for whom we borrowed a model rejected many old ideological presumptions and threw out many sacred cows before collapse.

The optimism and hysterical tone with which the public sector was launched have now given way to intense rage against it because it has let down the economy in respect of each of the promises and objectives tagged to it despite immense resources invested in it. Indeed, the main vicious circle revolves around the public sector, which takes about half the Plan investments.

When the public sector was started, it was given the following tasks: (a) to assume the strategic command of the economy,

A Decripit Public Sector

(b) to become the main instrument of India's industrialisation, (c) ultimately generate its own surpluses for further expansion, (d) to become a model employer, (e) to accelerate research and development, and to reduce concentration of economic power and income inequalities, (f) to set the pace of accelerating growth rate. It has fulfilled none of these great purposes. In fact, we find now that there is nothing public about it. The private big business leaves no opportunity in fleecing the public sector through a joint conspiracy between the business and the bureaucracy. It is certainly not even a well-knit sector; it is a string of enterprises run for and by the bureaucrats—imperial class—and frustrated professionals. Public sector along with public bureaucracy constitute an internal empire setting a seige against the nation.

About one lakh crores of rupees invested in the public sector in about less than 300 enterprises. Barring half a dozen enterprises, all of them have run in losses for years so much so that some of them have eaten up whole of their capital. The cost and time overruns have been as high as 100 to 200 per cent. Even the drum-beaters of the Nehru-Mahalonobis models would like it to be trimmed.

Market Failures

Various kinds of market failures, and some distributional objectives were used for shifting allocations from the private to the public sector as well for the justification of the price and other controls and regulations. The battered old theories of market imperfections and incompleteness were pressed into service even to justify intervention of all kinds without providing any empirical proof of the need for such intervention. In the heady days of planning in the fifties, there were some economists who insisted on not violating the basic premises of market equilibrium but the interventionists succeeded in giving preference to ideology over economic considerations.

Four arugments were adduced by the drum-beaters of the Nehru-Mahalonobis model. First, private sector would not be able to raise market finance for setting up big steel and other capital goods plants. This proposition was never put to test. Anyway, there was no justification for setting up an enterprise which could be set up by the private sector on that count. In fact, in view of the fiscal crisis and the consequences of attempting to raise budgetary resources for any new public sector plant only deepened the crisis.

The second market failure was seen in the prevalence of structural bottlenecks which did not permit equilibrium market clearance by free prices. Price controls were to disappear as the market structures developed. What actually happened was that price controls not merely created monopolies and market imperfections but also produced a vast rentier class.

The third market failure was seen right in the absence of certain kinds of markets such as rural markets, markets for savings, deposits, insurance and capital. All this had nothing to do with controls and regulations. Corrections of such failures fell in the realm of development. Government's policies have led to the creation of many kinds of markets but in a way that serious spatial distrotions emerged.

The fourth reason was the need for public sector or price distribution controls of certain public and essential goods and services. It is in this area that one finds both a justification and a great deal of confusion. Running public services inefficiently or subsidising them for the middle classes constituted the most perverse use of both the markets and the controls. What began as a market failure was compounded by the Government failure. The private sector enjoyed benefits from the public sector and the regulatory system, so much so that it developed vested interests in these regulations and shattered markets. The issue is well summed this by Dr. B.S. Minhas: "The nature of the state is also important in this context. Can we always take a public interest view of the Government? When a Government becomes a private trading post in which jobs, farms, policies and contracts gain, and rents of political and bureaucratic power are collected in most transactions, it is imprudent to be optimistic about the efficiency and efficacy of massive Government intervention in the production and provision of goods through public sector firms and agencies".

Therefore, seen in the light of the massive losses of the public sector causing tremendous fiscal damage, reducing money and capital for the private sector, creating monopolies and above all, massive corruption, it may have been more optimal to let market imperfections be part of development strategies rather than that of regulations and public sector monopolies and price discrimination. The irony is that the market operators have developed such a vested interest in that were created by Government controls that there are refusing to let the market go free.

Although I have no intention to suggest here Mahatma Gandhi's alternative approach for reducing the role both of the public sector and market for a different system as part of this analysis, a Gandhian warning is necessary. We are likely to face the worse of both the public sector and market mechanism because mainstream economic theory has failed to tackle the problem of wealth distribution without which no price is relevant particularly under pressures of globalisation.

The market failures and those of the public sector jointly made the latter grossly inefficient, corrupt and rendered it a great burden on the public exchequer. It has been a great drag on the Indian economy. An inefficient, financially losing and corrupt public sector is the biggest enemy of socialism with which it was wrongly equated. Yet, the proponents of the public sector like to join the losing ideological battle with those we would like to dismantle a part of the sector on economic goods. It is amazing that the Government never showed any reluctance to take over private sector mills, which have become sick, under political pressure.

Therefore, we see a strange sepectacle of a vicious circle which has become the mainstay of the bureaucrats, labour unions and the private sector. The dynamics of this circle runs something like this: with each new Five-Year Plan, this sector is ensured half of the new national savings: thus higher the national savings, the more of it is collected for public sector investments but the sector creates less and less surplus; the Government borrows or taxes more in order to feed the public sector investment. As the budgetary support to it is increased and as taxes could not finance it, the sector came to be financed from borrowing by the Government, thus burdening the exchequer further, through interest payments. In this way, the fiscal deficits were linked with public sector losses. Since the public sector largely produces infrastructural facilities for every lent class which benefited all the classes except the poor, it continued to enjoy legitimacy for its inefficiency and corruption. The more inefficient the sector the more money it got.

Nearly half the investment goes to the public sector without either accountability or economic rationality today, as it was 35 years ago, it is a sign of intellectual backwardness that the debate on public versus private sector still continues on old lines. The performance of the public sector has been so dismal that its further expansion simply undermines the budgetary resources. It is one thing to ask for improvement in the functioning of the public sector, it is and quite another to continue to allow this sector to put a serious strain on the economy and the fiscal system.

Altered Facts

The late Prof. Sukhamoy Chakravarty, the chief ideologue of the Nehru-Mahalonobis model, altered facts to make the public sector's poor performance linked to the decline in the growth rate of investment in that sector, which has always got and is still getting a lion's share in the investible resources. He contrasts the public sector investment growth of the sixties with a higher growth rate of the fifties. He conveniently ignored two facts: (1) that in the fifties, the private sector was not allowed to draw on investible resources, and (2) that, on an average, public sector investment increased at the rate of twenty per cent per annum during the eighties when the performance of the public sector was grossly poor. During the mid-seventies through eighties, the public sector's share drastically improved and the private sector's share declined. But, the capital-output ratio of the public sector kept on rising notwithstanding all the fluctuation. He ignores the basic fact that the Indian public sector was more an ideological sector than a business sector; it was corrupted by an uncritical defence of it by

A Decripit Public Sector

the ideologues, the bureaucrats and the politicians and even the private sector which preyed upon it. Of course, the public sector capital requirements were never fully met partly because the targets were kept too high or the resources were short. Anyway, this is more true of every other sector as well. The stupidity of preparing a plan on the basis of gross savings and not on net resources has not yet been given up.

According to Kalyan Raipuria, Advisor, Planning Commission, the public sector accounted for little over 45 per cent of aggregate investment in 1980-81, 51.3 per cent in 1984-85 in the base year of the Seventh Plan, and was 43.1 per cent in 1988-89. Public enterprises (departmental, DEs and non-departmental, NDEs) taken together broadly accounted for over three-fourths of public sector investment during these years. Thus, their share in aggregate investment works out to over one-third. The level of DEs' and NDEs' gross investment is estimated at Rs. 27,269 crore for 1987-88 out of an aggregate level of Rs. 72,134 crore. What is slurred over by the Nehruvian defenders of the public sector is that most enterprises have becomes inefficient, corrupted and parasitic that they are unable to generate adequate internal savings. Raipuria has shown that while investments have not declined in respect of gross savings, indicating resource generation in national accounting framework, the contribution of public sector has not been matching its draft on household savings. Public sector as a whole accounted for only about 16 per cent of aggregate domestic savings in 1980-81. This share came down to 10.5 per cent in 1987-88 and was 7.7 per cent in 1988-89 (E.P.W. 19.1.91).

Apart from what it might have done to itself, the public sector has strangled instead of promoting growth elsewhere, insofar as it failed to generate its own projected surplus, Plan after Plan. In real terms, the gross surplus ended up at less than half the projections, thus forcing the Government to drain its budgetary resources for this sector instead of putting the sector on notice. In net terms, over the decades supplies not only diminished but have turned negative. The illusion created was that the public sector was earning at least two or three per cent on the capital invested in only in gross terms. In real terms, i.e. net of depreciation, the net returns are not only negative, but the gap between the gross and the net has also widened thus demonstrating that depreciation funds went up because of heavy investment year by year.

The Nehruvians offer one major defence of the public sector, namely, that is promotes private sector investment in a long way. The elasticity is worked from regression analysis to be 0.73 (Bardan 1989). But, this is one of those cases either of the misuse of regression equation because of obvious multicoloniarity or an ideological bias to cover up the sins of both sectors. After all, if the public sector mainly helps the private sector at the cost of the rest of the society and ultimately produces losses, both sectors can be seen as having joined in a conspiracy. The public sector stands condemned, ironically, on ideological grounds. Thus, the complaint that public sector capital formation has declined for over ten per cent in the sixties and to relate it to latter decades constitute a peculiar disease of Nehruvian economists.

Probably, the worst role of the public sector has been its growing inefficiency and parasitism. Not only the returns on capital have declined and in some cases, even capital has been consumed, yet this sector has been given the privilege of maintaining its share of capital formation around 42 to 45 per cent continuously, while its share of gross domestic savings have continued to decline from about 23 per cent in the mid-seventies to less than 15 per cent in the mid-eithties.

A comparison with China may not be out of place because both India and China relied on the growth potential and savings of the modern sector. In China, the modern sector belonged to the public sector. However, whereas in China, public sector contributed about 25 per cent of the GNP in the form of savings that were transmitted to the fiscal budget. In India, the modern sector contributed than 4 per cent of savings and depended heavily not only on the savings of the households, which was consistent with property relations, but also on the budgetary deficits, an arrangement which is now in a very deep crisis. The so-called modern sector has failed India, as it were.

Parasitic Public Sector

Since the public sector has become parasitic rather than ensuring the commanding heights of the economy, there is little justification in continuing to waste resources on it, except for meeting the infrastructural needs at full cost pricing, security considerations and for some other critical objectives. Anyway, even within the original objectives set for it, the public sector expanded into areas in which it had neither business nor any social purpose, thus bringing about condemnation on itself. For instance, there is a general agreement even among the protagonists of the public sector that such enterprises as hotels, airlines and telephones should be immediately transferred to the private sector.

The most tricky point on the vicious circle was that of cost and time overruns and persistent dependence on budgetary resources at the rate which widened the gap between its need and its own internal resource mobilisation. Cost and time overruns have led to an escalation in investment costs by 70 to 80 per cent and their combined effect for generating capital waste ranged from 100 to 200 per cent. It does not need any profound analysis to show that delay in investments and cost and time overruns lower the growth rate, push up the capital-output ratio, lower productivity, all leading to undesirable price increases all along the line. In view of the fact that costs of infrastructural and intermediate goods dominate the public sector enterprises, the administered prices are not fixed in proportion to costs for fear of public criticism which results in the elimination of profits. Administered prices are allowed to go up to enlarge resources and profits which have no relation to costs. The enterprises neither care for the market nor for efficiency as they have no control over prices. The vicious circle gets more vicious because about one-third of the public sector costs are engineered within the structure itself as output of one enterprise is the input for several others. The vicious circle moves between increasing cost overruns, leading to an increase in prices which, in turn, causes further overruns and increase in prices. At each stage, the surplus of the public sector gets shrunk.

Public sector prices were loaded with too many objectives. Investment decisions were made independent of these prices or any other set of relative prices and profitability. Thus both price relationships and investment decisions were distorted. Worst of all, although a mixed economy, no clear dynamics of the command economy was ever worked out. No one could say with certainty whether and which prices approached the scarcity prices or played a dominant role in allocating resources in production, investment and corruption. A large part of public sector has to be privatised if resources invested in it are not to be wasted any further.

Extensive intervention by the state through physical and financial controls for the ostensible object of serving some social objectives achieved the opposite results. It has made the economy stagnate, generate inefficiency and high cost of the economy, and produce lower growth and high profits, thereby increasing inequalities. All that may have been corrected but for the fact that the state intervention ultimately created the most dysfunctional and yet the most powerful rentier class. It spread over from smuggling at the lower level to importers of capital goods at the top. The internal and external siege setters are found in this class.

The first task was to see that the policies, objectives and their social and economic effectiveness are assessed in advance. In fact, the contradictory character of economic effectiveness and social effectiveness was not recognised. In economics, whenever a goal is identified and possibly quantified, then effectiveness or principle of rationality can be seen in the minimisation of costs. This is not always easy as some costs and benefits defy measurement. Both cost and benefit are heterogenous and incommensurable. Therefore, the measurement of effectiveness requires clear indicators. The indicators of social effectiveness are quite different from those of economic effectiveness.

The variables that move on the circle are mutually reinforcing with different speeds. For instance, cost escalations and price increases of public enterprises either emanate from the same sources or re-inforce one another. For this, the well-known causes are:

The original estimates are prepared at original prices without taking into account the normal factor of inflation. Cost goes up because of inadequate project preparation and changes in the scope and many creeping errors. Changes in the expenditure on projects that occur at different phases are not foreseen as are changes taking place in the statutory levies during the period of installation of the project. Variations in the demand are neither calculated nor cared for because of state monopolies. Changes in the environmental factors and delay in the supply of inputs and equipment distort project costs and designs. In addition to these, the other important factors for the cost disruptions and miscalculations occur on account of inappropriate technology choices and thinning out of investment from non-economic considerations.

One can see that each of these factors which require enhanced cost escalations reinforce one another and, in turn, are reinforced by one or more other factors. In fact, the public sector became an entangled web of cost escalating circles within circles, all leading to one common result, namely, the frittering away of resources and increase in prices. If there is any sector which increases its prices and yet suffers a reduction in surpluses, it has to be dismantled unless it can show that it serves some larger purposes and that too in the shortrun.

Sebastian Morris, while analysing these factors, has come to the following conclusion, namely, that public sector's "problem has the characteristics of a vicious circle. The large linkages between public sector investments, delays, in a few enterprises get transmitted into other enterprises either as delays in supplies, or on the way of the accepted market demand being pushed into the future".

Briefly, the way the public sector has been run and its size and the proportion of resources it has appropriated when taken along with its inefficiencies and cost escalations the sector creates the most central vicious circle in the overall saving and investment relation. If saving and investment remain in dis-equilibrium all the time, it is largely because the public sector keeps draining the resources at a faster rate than the surpluses it generates for the national investable pool. This gap widens as the number of vicious circles, mentioned earlier, move at a faster pace through mutual reinforcement. It does not need very sophisticated calculations to prove that if the public sector absorbs more resources than it generates, and if half the investment takes place in this sector then the main vicious circle is between public sector losses, budgetary deficits, non-optimal investments, low productivity and then back to this sector's losses, economic slowdown, current and future, ending up in a variety of crises and disruptions in the economy.

Public and Private Sector

The private corporate sector had a rich fund of managerial talent but it was short of equity and other finance for growth. On the other hand, the public sector had the finance, but a disappointing profit record, attributable mainly to lack of operating efficiency, market orientation and clarity of goal. Administrative delays and a bureaucratic rather than commercial style of management had affected the public sector performance. The crux of the problem was how to promote the public sector's managerial efficiency. The answer perhaps lay in setting up a corporate structure where management inputs came not from Government but from private sector. Even as things stood, many so-called private sector companies had 10 to 30 per cent public sector shareholding, and over the next 10 to 20 years; for some of the large non-FERA companies this percentage would rise significantly, perhaps to over 50 per cent. The distinction between the public and private companies was thus likely to get blurred in due course of time. Against this background, it would have been worthwhile to build new plants and factories with government resources and private sector management.

Thus the most striking siege around the economy is laid through an indifferent and money guzzling sector public sector units by the joint operation of managers, bureaucracy and private sector. The private sector both fleeced and undermined the public sector but the public sector has suffered even more at the hands of the bureaucracy. Indeed, the professional managers were so much constrained by the overlordship and interference of bureaucracy which knew next nothing about business management, that they ultimately provoked the latter to ruin the public sector. The performance and the very existence of public sector was totally jeopardised. One way to cover this damaging situation and fraud was the signing of the so-called memorandum of understanding between the Government and public sector undertakings with the ostensible objective of giving autonomy with accountability to the public sector. About 26 companies covering about 75 per cent of the total investment made in the public sector and 80 per cent of the turnover have signed such phoney memoranda.

The professional managers make no bones about their still being strangled by the bureaucracy notwithstanding all the changes made in the format of the memoranda. The chief executives of the public sector undertakings regard the new format as yet another instrument of unequal relations between Government and PSUs. The problem is that MOU involves two unequal treaties. Both the company and the Government are required to adhere to specified targets or schedules but practically every chief executive has complained that their parent Ministries do not clear projects as per MOU schedule and yet the companies can be hauled up if they fall short of their goals.

In conclusion large part of the public sector has run out of its utility. It is draining and wasting national resources. It is caught in a vicious circle which is one of the various circles on which the economy is moving. It has been put in strait jacket by vested interests. But quite tragically there is little possibility of reform, what to say of dismantling this sector though without it is the economy will remain under siege. The defunct public sector sustains and in turn is restored by the equally defunct ruling elite.

Chapter 8

The Ever Rising and Corrupting Middle Class

The internal seige against the economy, indeed against the entire nation, is laid by India's ruling class which has become insensitive, cruel, unaccountable, irresponsible, corrupt, intellectually bankrupt and very much integrated with the old colonial West and increasingly alienated from its own people. The middle class is a critical part of this ruling class or elite.

There is a theoretical problem. What is a middle class? There is no consensus on it. For a working definition in the dynamics of Indian social structure, it is a class in the middle of the stratification made on the basis of anyone or more of these powers: income, wealth, social status besides being on the top of the knowledge power. It is a collective product of the society's dynamics of hundreds of internally conflicting and co-operating relations. Notwithstanding the Marxist nihilism, this class has bulged, in the process invalidating Marxist class polarisations.

Without going too much into details, we define the Indian middle class to include (i) the small businesses both manufacturing and trading, (ii) the middle peasantry, (iii) the entrepreneurial classes, (iv) a large chunk of the bureaucracy, (v) the professional classes, (vi) a variety of groups performing many overlapping functions in the informal sector but having incomes comparable to those of the other aforementioned categories. and (vii) the political elite who may have earned or unearned incomes. Only the first three categories comprise the producing classes. It is difficult to give a more precise definition as each has a different role but income-wise, all these can be lumped together. It is even more difficult to give their exact number, but it is estimated around 100 to 150 million. Since the class above the middle class cannot be more than one per cent of the population, the two together extract the economic surplus from the remaining 750 million people by way of surplus labour, rents, corruption and inflation.

There is also no ideological self-definition of the middle class. That is why there is a resistance to speaking of middle class politics or middle class economics. However, there is no movement of significance in which the middle class is not involved, whether a particular movement is radical or conservative. In a somewhat non-Marxist fashion, the middle class co-operates and conflicts with the *bourgeoise* and resists being proletarianized. *It says put and that is why it is called the ever rising middle class.*

The Indian socio-economic development model, despite all the slogans of socialism, has essentially been one of mixed economy capitalist welfarism within a democratic polity. The general thrust of the social policy has been that since economic development will create new inequalities and disparities, their impact must be moderated by a vast expansion of such facilities as will create equality of opportunity. It has been proved beyond a shadow of doubt that a system based on equality of opportunity has helped to expand the middle class. The vast expansion of educational and health facilities undertaken was appropriated by this class. India is one country where expansion of public sector bureaucracy has been most rapid. The very nature of the mixed economy helped the creation of a vast class of middlemen between the Government and the private sector. Advertising and other commercial services, fine arts and not so fine arts and research and other scientific institutions further multiplied the middle class.

The bureaucratic dominance of our planning and economic development helped in the growth of the middle class. This trend was most prominent in the services sector but its base was laid in industry. The employment growth of wage earners in the organised industry was overtaken by the salary earners. Whereas agriculture and industry grew at uneven rates, the tertiary sector's growth was above the average growth of GNP.

The growing unrest and violence in the country is not independent of the changing class constellation. The middle class generally respects the concentration of the tools of political power. When less privileged groups exercise violence to get a response from the political system, the middle class resists it for obvious reasons. But when its own position is threatened, the middle class not only does not disapprove of violence but itself resorts to it. This is often done in the name of national unity and the so-called nation-state or national security state.

The Farce

The farce of creating a new Indian middle class has ended in several national tragedies and civil wars, all of which were led by the same class. Every time there is an economic crisis, economists seek shelter under solid base of 150 million persons of middle class which constitutes one of the largest markets in the world and is also the guarantor of stability and growth.

The lower classes have always struggled to rise to the level of the middle classes both in economic and cultural status but their entry has been very meagre. The middle class multiplies itself very fast to leave little space for others. Yet, notwithstanding all its privileges, the Indian middle class is given to disillusion and despair.

If one goes by the number Income Tax payers who have cars, refrigerators, televisions sets and modern houses of more than 200 sq. yards plinth area the number would be rather small. If one extends the definition of new people who live in *pucca* houses of any kind and who own scooters, the number may be 25 to 30 million families. What is interesting is that the top two decile of the population own three-fourths of the wealth and income. just because the rural middle class may not have some of these things, they cannot be excluded from the definition. The total number may be about 200 million people at upper limit.

This class has several very prominent characteristics. (1) contrary

The Ever Rising and Corrupting Middle Class

to the Marxist view, the class is an ever expanding class and levels off the pressures produced by the poor working classes at one end and the *bourgeoisie* at the other; (2) it is the mainstay of national savings as well as the source of the most conspicuous consumption, (3) unlike its counterparts in the developed world the Indian middle class enjoys personal services such as domestic and street service at cheap rates, creating pseudo employment at low wages; (4) a section of the class contributes to the ruling bureaucracy which protects the interests of the entire elite class; (5) this class self-defines and self-legitimises though not ideologically, but through vast Government subsidies for education and other services at the cost of other classes; (6) it is the class of which a majority is parasitic and is pitted against the producing classes; (7) most damaging role of this class is its intellectual subservience to Western culture through its integration with outside than with inside forces. These characteristics should leave no one in doubt that they all add up to the main battle force for internal and external siege against India. Paradoxically, this class is more angry than poverty stricken people because of its double alienation, alienation from within itself and the alienation from the masses. That is why as it grows in numbers and wealth, it keeps moving into wide waste lands of nihilism.

Without fiscal support, this class could not have assumed a commanding position. There has been a deliberately designed policy of giving several measures of fiscal support to the middle class. Whereas in 1950-51 the personal income tax accounted for 21.37 per cent of total revenue, by 1987-88 this ratio fell to around 5 per cent. The Government revenues and expenditure increased several times during this period, and thus the burden was shifted to other classes through taxes. The inequitous character of this shift clearly suggests that the growth of middle class, even if justified on other grounds, was fiscally highly undesirable. The main instrument was a series of fiscal concessions given to personal income-tax payers. It has been shown by Dr Nayak and Dr Aggarwal (EPW 9.7 1989) that the prevalent exemption limit is very high in keeping with the international comparison. Instead of lowering the limit, there has been a persistent demand from the middle class to abolish the income tax altogether.

The fiscal system provides four types of tax concessions i.e. (a) exemptions under various sub-classes of Section 10 of the Income Tax Act; (b) tax concessions of immediate deduction to long term financial capital gains under Section 54 and (c) National Savings schemes which allow immediate deduction but require adding back amounts withdrawn to the income of the year in which withdrawal is made; (d) exemption on gift tax, (e) even under a proclaimed progressive income tax, excises and other fiscal imposts allow the middle class to expand both in size and economic power. (f) The concessions on wealth tax do the real trick. The exemption is adding up to five million rupees in financial assets such as saving bank accounts, provident funds, capital investment bonds, public undertaking bonds and same bank deposits. (g) In addition, the elimination of self-occupied houses and expensive interior decorations from the ambit of wealth tax has gone a long way to strengthening the middle class.

If a person has an income of a lakh of rupees, the marginal rate of tax is 50 per cent. But after standard and other tax deductions, the ratio comes down to 29 per cent. However, there are so many other tax shelters available through the concessions mentioned above that one need not pay any tax at all. Ironically, in the first instance, concessions are built to reduce the tax liability to absurdly low limits which is then followed by the demand to abolish the income tax altogether.

In a study by Das Gupta about the advantages to income earners of 26 types, it is shown how the Government's various financial instruments of borrowing are destroying the tax system and its incidence. The purpose of using these assets is largely to encourage household savings which now constitute 80 per cent of total domestic savings. These instruments are used as tax shelters. Savings will surely decline without these concessions. Das Gupta has come to three conclusions which demonstrate that all of them help the middle classes and high income earners.

First, financial instruments of public borrowing reduce drastically the tax liability of the buyers of these assets. The Government trades off borrowing for taxes and thereby strengthens the financial base of middle and upper classes. Second, the ranking of assets after income tax creates a system quite different from the nominal system. Finally, the budgetary implications are that the fiscal system simultaneously encourages savings as well as consumption of the middle class by the transfer of funds from the Government that take place through the difference between the Government discount rate and the effective rate of return.

Since 1978-79, when the gross domestic savings as percentage of GDP reached the figure of 23.2 per cent, the savings rate declined and has remained stuck at around 21 per cent. But whereas household savings have gone up, there has been a persistent dis-saving in net terms, net of depreciation, in the public sector. In the private corporate sector, savings have been stagnant. But more significantly, during the last decade, tax provision as percentage of sales has declined from 3.8 per cent to 1.8 per cent. This was because of much greater reliance placed on borrowed money rather than on equity, which shifted the corporate sector's saving potential to others. A critical aspect of household savings has been the increase in the ratio of financial to physical assets. The supply and demand base of middle class was expanded in this way. This fact has to be read along with decline in the ratio of commodity production in the GDP.

The Consumption Pattern

Besides, the consumption pattern has shown a big shift away from non-durable goods to durable goods i.e., the consumption basket has tilted towards the higher income groups. During the eighties, the growth of industrial durable goods has ranged between 18 to 40 per cent as against 5 per cent of the non-durable goods. Besides consumer durables, three other goods and services which reflect the shift in the pattern of consumption in favour of the middle class. According to S.L. Shetty, these are (i) some capital goods industries (computer systems and office equipment) which may be subserving the interests of the high-income bracket consumers, (ii) a substantial amount of clandestine imports of consumer durables and other middle class consumption goods (cigarettes and liquor) which do not get reflected in the official production and consumption figures; (iii) a large set of service industries (such as hotels and trade which also subserve the needs of middle and upper classes).

In addition, the acceleration of the rate of inflation transferred incomes from the poor to the rich and to the asset holders from the non-holders of assets. The upward push in the interest rates, combined with fiscal concessions as well as the relative shift in financing of Government expenditure from taxation to borrowing, helped the holders of assets the middle class.

Shetty further states that the policy of high effective yield rates with substantial fiscal concessions has begun to be counter-productive. Studies have shown that the total household savings hardly respond to high effective yield rates. But, something else happened. With significantly high yield rates, particularly on non-marketable securities, the wealth holdings of high and even middle income households get augmented which, in turn, produce an effect on consumption and saving in that high effective yield rates tend to encourage luxury and semi-luxury consumption. Even if such households continue to undertake sufficient financial savings so as to take advantage of fiscal privileges, their consumption is also augmented in relation to the consumption of the rest of the community. Consequently, no increase in aggregate domestic saving takes place. Shetty says, "The community at large suffers in two ways: first, a redistribution of consumption takes place in favour of a small segment of society at the higher end of the income scale and in favour of luxury and conspicuous consumption; the secondly, it contributes to an investment and production structure that is socially less desirable. On the whole, development priorities are distorted". In this way, the paradox of high consumption and high savings of the middle and upper classes is explained. The middle class could not have maintained its high savings and consumption without the increase of the interest rates offered by public borrowers. With nearly double the interest rate now as against that of two decades, ago, the middle class incomes have been protected under deficit induced inflation and the strong money illusion. The conventional economic theory has broken down. The whole system is seen by the middle class as its private affair.

What has happened in stock exchanges in the last few years offers a great lesson in how the ruling class manipulates the system entirely for its own sake even when, as a consequence, the nation's wealth-creating opportunities are undermined. When the Government financial institutions intervene to encourage speculation instead of curbing it, the results cannot be beneficial in the long run except for the speculators. The maturity and rapacity of the middle class is buttressed by the maturity and rapacity of the speculators. All this is easily seen in the very high average price-earning ratios of the 'A' category scrips which set the trend of prices at the stock exchanges.

The reasons for soaring high stock exchange prices are: rise in industrial production and profits; the entry of the mutual funds of several thousand crores of rupes as addition to investible funds; the massive involvement of LIC, UTI and GIC in the buying and selling of scrips, spurt in the NRI funds and above all, chanelling of black money into the share market.

The bureaucratic middle class in alliance with big business allows a lot of concessions to the small scale business. If middle class business houses are hurt, the whole class is hurt. But how to meet the danger of big fish and the still bigger fish of multinationals strangling small fish i.e. small business? Through a variety of fiscal concessions, differential interest policies, preferential purchases by Government, the small units are preserved. But many have gone sick and are retained with bank money and corrupt practices. This is an internal arrangement within the ruling class, the burden of which falls on the poor. The poor pay for the internal imbalances between different components of the ruling class.

A recent study has shown that nearly 20 thousand crores of rupees are outstanding bad debts, most of which has gone to trade and industry i.e. the middle or upper classes. The powerful rural sections did not want to be left behind. Through the growing political power, they extracted a scheme of loan waivers for about 10,000 crores of rupees. The urban and rural middle class is extracting a fat share of credit and lets it be mismanaged. A large chunk of accelerated growth of the middle class is made possible through the ravaging of the financial institutions.

Industrial Sickness

It is often claimed by the business class that industrial sickness is due to the absence of modernisation and lack of funds to undertake that modernisation. Lack of funds is attributed to high basic rates, disallowance of certain expenditure which increased the effective tax rate, depreciation and old-fashioned structure that is inconsistent with the problem of rapid obsolesence, technological changes, product differentiation and inflation. The Government responds positively in that high taxation is counter-balanced by a variety of incentives. The Government is not oblivious of the fact that these policies are heavily weighted in favour of the big against the small. For instance, the Directorate of Special Investigation, Ministry of Finance, has observed, (as quoted in the 143rd Public Accounts Committee) that "most of the beneficiary companies of all the provisions relating to incentives/concessions are owned by monopoly houses. Only fully developed companies with a large capital base and availability of substantial institutional finance can derive most of the benefits. Only monopoly houses own or float such developing and developed companies with vast inter-corporate resources and availability of institutional finance. In a nutshell, the large companies pay a lower tax rate and the vast majority of the industrial units pay a higher rate of tax because they are unable to avail the incentives for reducing their tax rates." The latter are then given other concessions. All these are problems of the dynamics of conflict between the middle class and the *bourgeoisie* which are sorted out by policies that protect both under the smokescreen of progressive taxation.

Another powerful trend towards the making of the middle class is the unusually large creation of the most durable of all durable assets, i.e., residential housing. Construction takes about 50 per cent of investment for the business, Government and households. During the eighties, the share of construction in the Government did not change but for the household it declined. In fact, the share of construction went down for all sectors combined, except for the Government sector in which it increased. Household savings in physical assets consist of construction and machinery (and stocks of both). However, a study by Roy and Sen (E.P.W. 20.4.91) shows that residential construction as a proportion of GDP rose quite significantly from 1.9 per cent in 1970-71 to 3.9 per cent in 1984-85. Since the rate of household construction as a whole has declined, it follows that the decline has taken place in the non-residential buildings. "Presumably this construction of workshop and office space (is) for small industry and trade... To sum up about construction, there has been indeed a boom but only in residential housing and probably also in corporate sector's non-residential construction".

It is seldom realised that the new middle class is being encouraged by several other policies which are kept out of our analysis. For instance, the trade union power, massive growth of public sector bureaucracy and the rise of the white collar class which now give a big lead to the middle class. But these are not isolated factors. Both the Government and the large companies prefer to confine themselves to this market than risk the emergence of a competitive capitalism for the expansion of the home market.

For a long time, the urban middle class kept the money and political power largely to itself. The power is shifting now. There is a new rising middle class in the rural areas about the role of which our Nehruvian economists and political scientists have been less certain. India is fast moving into a situation where the rise of a peasant power is making its mark both on Government policy and industrial structure. That is how the middle class got bloated and it thereby defeated the assumptions of the Nehru-Mahalonobis model. A part of the subsidies are given as a compensation for the adverse terms of trade between agriculture and industry.

Once upon a time, the middle class as inheritors of both the nationalist and imperialist traditions exuded great confidence in India's future. This class backed Pandit Nehru and his confused model against the Gandhian model and Pandit Nehru in turn buttressed this class both by retaining the old ICS and the colonial education system. An extra edge was given to the class by Pandit

Nehru's fear of and struggle against the Indian *bourgeoise* which had yet to come on its own. The massive increase in the size of the bureaucracy and an army of Left intellectuals completed the job. There were no ideological trappings given to this class until the Nehruvian model of development and foreign policy failed. There came the ideological role of this class. The Left intellectuals who were getting discredited were briefly rescued by Kalecki and Joan Robinson and by digging up Marx of the *Eighteenth Brumaire*. A new theory of the Intermediate Class was put into currency to justify the state being taken over by the middle class and to cover up all its ugly manifestations. No one talks any more of the intermediate class as it has become totally parasitic, bloated and farcical. Yet, its hold is complete.

The greatest transformation that has taken place in India has been most perverse. As the middle class grew and grew very fast, its functions turned from positive to parasitic except in producing components. The cultural arrogance of this class produced the ugliest divide between this class and the masses, bringing in its wake or revealing its inability to meet the challenge of economic instability, social discontent, political turmoil and new mini wars. Above all, this class, along with the big business, wilfully integrated itself, as a subsidiary, with global economic knowledge and order and thus created two Indias. The new slogan is globalisation. Ironically this class failed the country in areas in which at least it was expected to contribute most i.e. managerial, organisational and technological and educational fields. The more it failed, the more parasitic it became. The relationship between this class and the Indian state was one of mutual reinforcement in parasitism, oppression and subversion of development needs. Every component of the middle class has become an internal empire with total moral nullity, particularly the bureaucracy and the English-language educated parts.

CHAPTER 9

The Rentier State

It is most surprising that no economist before or after A.O. Krueger wrote her piece on the retier class in 1974 has researched or analysed the growing strength and role of the rentier class. If nearly one third of the national income is generated through black income, tax concealment and artificial scarcities, surely this was too important a subject to be left out. To me, three reasons are obvious.

First, in the economic theories developed by the West, there is no significant place given to the study of the rentier class because such a class was of very minor significance. In a relatively free market and open commercialisation, the chances of the growth of rentier class remains slim. The laws are faithfully implemented to a large extent and conformed not only by business but also by those who job it is to conform to laws. The members of the Indian ruling class are never punished for violation of laws and indeed they take pride in manipulating laws as well as scarcities. Second, the educated middle class whose education is financed through high subsidisation and its burden falls on the poor is itself partially rentier in nature. It is reluctant to discuss the role of the rentier class in Indian economy. Third, in the Marxist model of development, rentiers either do not exist or are expected to disappear as capitalist development matures.

None of these reasons are good enough for ignoring this class as it is the most dangerous seize-maker. Indeed, other siege makers could not flourish but for this class. Notwithstanding all the objections to recent financial reforms, one good may come out of it all. If some of these controls and regulations which created scarcity rents are removed, the size of the rentier class may be reduced.

The most powerful class in India is the rentier class. This class has caused misallocation of resources, distorted relative prices, generated non-development inequalities, caused a shift from investment to consumption, encouraged corruption and above all, made a mockery of planners and policymakers by slowing down the growth rate itself. It is this class which sustains the dualism of two Indias.

What is the character of this class and why is it the most important? Before we go into this question, it must be clearly understood that those who earn economic rent on land and property do not fall in this category, although under conditions of rent control and accute housing shortage, economic rent remains a matter of academic interest. It is not a problem of distribution of economic rent between two classes of owners and tenants. However, those who earn from land speculation or land grabbing and land reservations under conditions of control belong to the rentier class as defined here. This class is system-oriented. In laisse faire economy of totally free competition, there will be no rentier class as the market will clear excess demand or excess supply in both commodity and factor markets. The long run equillibrium will be the sum of a series of short-run adjustments openly made in the open economy.

In reality there is no such economy. All economics are subject to government intervention. But the vital distinction is between those economies in which the government intervention aims at removing the distortions of the market mechanism and those in which competition is hampered by excessive controls and regulations. Of course, all states intervene for welfare activities. There are other complexities and purposes of government controls but the main objective of optimum market rationality remains the dominant consideration.

A Rent Seeking Society

India is not only a planned and a highly controlled economy but also a rent seeking society, precisely as a consequence of the Government controlled sector having assumed sufficient autonomy and power even to erode planning. Let me give some specific examples. Under controls and regulations, i.e. the licence permit raj, those who are able to get licences by using whatever methods, those who are allotted foreign exchange in situations of overvalued rupee, those who act as middlemen to get machines, power, railway wagons and raw materials out of turn by paying bribes are the rentiers. This class is not entrepreneurial but parasitic. If a proper breakdown of the entire income generated in black economy were available, it would be found that most of the income belongs to the rentier class. If, as estimated conservatively, nearly one hundred thousand crores rupees of income are generated in black economy, most of which goes to the rentiers, then this class is more powerful than the genuine industrialist, trading or working classes. It is difficult to give a numerical estimate of the rentier class but, by and large, it must belong to the top 20 per cent of the population which take away, legally and illegally, about three-quarters of the incomes and wealth.

This vast rentier class is the product of the discretionary powers of the controlled economy, the arbitrary power of the bureaucrats and the unholy alliances between political, business and bureaucratic elite. Under this class configuration those who claim to plan are deceiving the nation and also themselves. The Indian economy has been subjected to multiple regulations and controls for the ostensible objective of planned development, sectoral balances, channelling of funds to priority sectors, price stability, equilibrium, balance of payments and what not. In reality, the Indian regulatory system has become a self-perpetuating monster. The multitudes of rules, procedures and laws have created such an autonomy of power outside the framework of rational policy that it is difficult to prove whether or not a particular control is necessary at all.

Not only the setting up of industries, but also their maintenance expansion and growth along with price, profits and wages remain under one control or the other. Credits, interest rates, capital issues, exports, imports, almost every imaginable kind of economic activity is subject to one or the other regulatory law. Regulations lead to a great deal of subordinate legislation that confers a wide range of powers on the bureaucracy which too has become more and more rentier in character. Often, the existing controls breed a new kind of controls which gives each beneficiary the justification to strengthen the laybrinth. Ironically, the size of the rentier class and controls have strengthened the large business houses as against the small units though the objective was to reduce the power of these houses. Industrial policy aimed at reducing concentration of power but the mediation by the rentiers produced results quite the contrary.

Three arguments were used in defence of massive regulations and controls of industry which have also helped to generate the rentier class. First, the market mechanism will not yield the kind of decisions that are needed to achieve national objectives. Second, economic justice along with other economic objectives will be best achieved by the state creating the commanding heights of economy which were equated with both the public sector and regulatory structure. Third, controls were justified for purposes of rational allocations of resources and reduction of scarcities, which ironically produced new scarcities. No restrictions were imposed on controls; nor was any mechanism provided for the liquidation of unworkable or harmful controls. Thus, along with the desired effects even more undesirable effects were produced, creating vested interests and power centres in the way. No wonder, the power hungry planners and bureaucrats are unwilling to see that their policies have produced neither justice nor equality nor even rationality. Controls were equated with planning and became a sacred cow.

The rentier class like the other classes is internally competitive, but quite paradoxically, competitive rent seeking is the product of a non-competitive economic system which creates divergence between the private and social costs of certain activities. When quantity restrictions and constraints are imposed on domestic production or imports through licensing, a class is born which is largely parasitic, unproductive and has vested interests in

The Rentier State

multiplying controls. The distinction between genuine planning and unnecessary controls is truly drawn by the rentier class.

Several other policies and sources have further reinforced the rentier class. First, the most important is the tax system. About 87 per cent India's tax revenue comes from indirect taxes, and the remaining from direct taxes. The situation in the developed countries is exactly the reverse. Year after year as a consequence of reckless increase in the Government's expenditure new taxes were raised in the form of excise and custom duties. At one time, the rate of direct tax had crosses the limit of 100 per cent. The rising expenditure could be met only from indirect taxes. High direct tax rates led to enormous evasion and avoidance of taxes. This was the begining of the creation of another section of the rentier class which got hold of the undeclared production and transferred the cost to the declared production figures. Although the direct tax rate was reduced to 60 per cent in 1977 and since then it remained more or less the same but the continuous increase of plan and non-plan expenditure forced the Government to rely even more on indirect taxes.

These fiscal policies became the instrument of enlargement of the rentier class. For instance, through a variety of inside sources and leakages, the business community came to know in advance of the indirect taxes likely to be imposed. They withheld supplies from the market, creating a double stimulus, legal and illegal, towards generation of incomes which created another chunk of the rentier classes. Even if they do not have inside information, in a situation of steadily rising excise duties stockholders enjoy enormous rents on stocks already hoarded. Indeed, the distinction between a businessman and a rentier disappears.

The Annual Budget

The annual budget of the Government is presented on a fixed day of the year, with straight forward imposition of new excise duties. In a fiscal system dominated by indirect taxes, no fiscal principle justifies a fixed date in a year or year by year increases. Consequent upon a new excise duty, stockholders, whether traders or producers, are given millions of rupees of incomes, as prices of goods go up.

Second, the licensing system, foreign exchange and import controls became the most critical factors generating the rentier class. All this began during the war when war-time controls were imposed. Other countries removed them after the war but India not only retained them but made them more elaborate and complicated in the name of planning. As mentioned earlier, licensing was meant for controls of resources but an illegal market for licensing was created which produced competition for rents. One need not be an entrepreneur if one had a licence; the licence could be sold at a price.

A licence holder was not always the direct user. He often sold it to others and thus earned rent on the licences. But in most cases, where the importer was a direct user, he also enjoyed a large rent. The licence were presumably allocated in proportion to firms' capacities. Consequently, investment in additional physical plant conferred upon the investors higher receipts of import licences against those who did not get the licence. All producers did not get the same advantage. The difference between the actual cost of import and its market price was the first stage of rent creation. At the next stage came the scramble for acquiring new import licences through the link-up of politicians and bureaucrats, thus enlarging the class that fattened on rent-seeking. Long ago, the Santhanam Committee had estimated that import licences were worth 100 to 500 per cent of their fact value.

Third, a class of bureaucrats which shaped and managed controls developed vested interests and openly thrived on bribes which formed a part of the rental income of the licence holders. In this way, the bureaucrats too became a part of the rentier class through a transfer payment system. It is not surprising that the I.A.S. became the most prized job becuase the remuneration was calculated both in terms of salaries as well as bribes. Beside, a whole class of intermediaries outside the administration cropped up to act as middlemen for rent-seeking. A vast network of so-called consultants exists to do this job. The Bofors deal has exposed the antinational role of this class.

Fourth, the tie-up of business rentiers with the bureaucracy

The Rentier State

became quite sinister as both created and shared transfer payments in a highly complex entanglement with the underworld. Apart from smuggling, the scarcity of essential imports and Government controlled services contributes to the size and wealth of the rentier underworld significantly oppreases the other India. This can be shown by giving glaring examples of large operators. Railway wagons are in real or assumed shortage. To get a wagon in time, even for exports, bribe has to be paid. In areas of power shortage, some get power regularly if they can pay a bribe. For a long time, a car buyer had to pay excess over the garage price. Land reservations and ceilings in urban areas have become a multi-billion rupee scandal. Even capitation fee has become quite common to get a seat in a preferred educational institution.

Creation of the Rentier Class

Indeed all controlled commodities, such as cement, sugar, cars, scooters and food are the sources for the creation of the rentier class. Examples can be mutliplied. Every segment of economy now has a component of the rentiers. Competition to get into Government service, entry of a firm into business, struggle for being posted in a job which enables earnings over and above the salaries and to get a house alloted in order to sell it are examples of rent-seeking activities.

Probably the most critical aspect is the existence of a vast financial market outside the organised banking system. This informal market both finances and is financed by the rentier class. The interest rate differentials between the bank rate and *bazar* bill or *hundi* rate can give some idea of the power of the rentiers financial transactions. This informal market often makes the Reserve Bank's monetary policies look foolish.

It is not possible to spell out how the rentier class can be eliminated. But a few steps are absolutely necessary. First, and the most important, Government expenditure, plan or non-plan, must be statutarily reduced and limited to a fixed proportion of GNP. Second, all licences, for imports or setting up of industries, should be auctioned so that if rent creation has become inescapable due to scarcities, open competition for rents would get shown into legal market income.

This ratio of indirect to total tax revenue should be gradually reduced. Along with this, indirect taxes should be changed not yearly but after three years at least. Fourth, tax proposals should not be announced on a single dy. Fifth, since all markets have a rent generating component, there should be an elaborate list of fiscal imposts on windfall gains. Sixth, the inefficient public sector units must be dismantled. Finally, there must be an annual review of controls and a system of self-liquidating controls established.

As mentioned earlier, the rentier class renders Marxist analysis redundant. Mahatma Gandhi once told the Indian Communists that he was not averse to class analysis but the main distinction to be made was not the one between the *bourgeoise* and the working class, but between all the producing classes on the one hand and the parasitic classes, including the state, on the other.

CHAPTER 10

Liberalisers Without Liberalisation

Liberalisation has become an absolute necessity because controls and regulations that have been piling up are blocking initiative, encouraging corruption, reducing efficiency and causing misallocation of resources. Although, every body shouts about deregulation and liberalisation, specific demands depend on sectional interests. Neither industry nor the business community is homogenous. Different interests have benefitted or lost in different degrees from controls. Therefore, it is natural that with each section's response to new Government policy should be different from that of the other. Besides, the various interests represented in the Government itself do not hold similar views. What is given with one hand is taken away by the other. Liberalisation itself may be external, internal, sectoral and spatial.

There are producers and users of imports. By and large, the producers will have to compete and the latter only to benefit from liberalisation of imports. Similarly, there are producers who are tied to multinational companies and will have one kind of response as against those which are largely Indian companies. The former will always have some advantage over the latter. In between, there are a number of FERA companies as well as Indian companies with foreign collaboration which are categories by themselves. Some will benefit, from external liberalisation, others will not.

The same applies to internal liberalisation, but for different reasons. For instance, internal liberalisation yields enormous advantages to large industrial units as against the small and the tiny. Just as even some large industrial units are afraid of external liberalisation because of the power of the multinationals, the small units are afraid of the big units within the country.

Probably, the most difficult challenge would be to those who are trying to innovate domestic technology. External liberalisation can result in liberal import of technology which is a double-edged weapon. It can promote technological development by forcing people to upgrade their technologies and at the same time can kill innovating entrepreneurs who cannot compete with the foreign technologies. Besides, those who have preferred to import second rate technologies through collaboration can subvert a best conceived policy.

Therefore, liberalisation has to be a package which would include diverse elements in order to meet the complexity of the problem. Any blanket liberalisation policies, internal or external, can be counter-productive. It is universally recognised that without domestic liberalisation, external liberalisation would be most dangerous. Unfortunately, the Government has gone the other way around under pressure from the bureaucracy which may be forced to yield on external front but will never yield on domestic controls and regulations. In this, the question of economic rationality is of much less importance; the real question is whether the collective bureaucratism or the bureaucratic and the command structure of the state economy can be broken. Given the weakness of the political system and the enormous power the bureaucracy has assumed, it is most unlikely that domestic liberalisation will come first.

No Secret

It is no secret that the business community enjoyed enormous benefits from controls and regulations even when their own initiative and investment programmes were constrained. The business did not like the controls their political masters imposed on them for ideological or other non-economic reasons. But the system of controls allowed the business community to have sheltered markets which gave them enormous scope for making profits, legal or illegal, which they shared with the politicians and the bureaucrats. So long as controls ensured the sheltered market, the business did not feel bound to improve technologies as the old or imported technology could do the job in hand. Indeed, if the business community opted for controls, it was a trade-off for the demand for nationalisation from the left-wing of the political parties.

Let me give some samples of the reaction from the industrialists to the Government's new policy of devaluation and liberalisation. These are taken from a daily newspaper of July 15th, 1991:

R.P Goenka, chairman, RPG Enterprises, maintained that the domestic industry was not in a position to face international competition overnight, said that the Government should phase out the implementation of the liberalised import regime over a three-year period.

While welcoming the objective of reducing tariffs, Goenka warned that unless it was done gradually, the domestic industry would suffer a major setback.

Rajan Nanda, vice-chairman and managing director, Escorts Limited, would like the entire structural reforms to be implemented in one go. This is mainly on account of the fear that a phased implementation would defeat the objective with the various anti-liberalisation lobbies coming to the fore.

Maintaining that the domestic industry, which had for long remained insulated would inevitably resist a change. He firmly believed that the economy which had been "driven for long" needed to become a part of the global economic framework. And, whatever needs to be done towards this end, "should be done forthwith".

Ashwini Kumar Puri, chairman and managing director, Mohan Exports, said that while the country's exports were expected to dwindle following the devaluation as also the withdrawal of incentives and subsidies, the immediate liberalisation of imports would prove fatal to the economy. While on one hand, the cost of manufactured goods would go up making exports even more difficult, on the other the drain on foreign exchange would only aggravate the situation. "The Government should see the impact of liberalisation that has already taken place on the overall export performance for two to three years before considering further liberalisation of imports. It could even lower the tariff to 50 per cent rather than to peg it at 100 per cent," he added.

M.V. Arunachalam, chairman, Tube Investment of India Limited, felt that the proposal to reduce the tariff to 100 per cent was a welcome step only if it was part of the first step of the phased restructuring programme. In the second stage, the duties should be below 50 per cent, "if India is to become an integral part of the global set-up and progressively move towards a floating exchange rage regime."

Favouring a similar reduction in the import duties on capital goods, Arunachalam said that the Government would be able to offset a part of the revenue loss with higher exports under a REP regime.

K.N. Modi, chairman, Modi Spinning and Weaving Mills Company Limited, said that he would like to see the whole package, including the changes in MRTP and FERA laws. According to him this should do away with a lot of fat accumulated over the years in a protective environment, and face international competition. This will improve its competitive skills," he said.

Onkar S. Kanwar, managing director, Apollo Tyres Limited, who firmly believed that the "era of the survival of the fittest" had arrived, favoured an immediate introduction of a tariff rate of even less than 100 per cent to begin with.

The country was virtually starved to capital goods, and as such there was a strong case for making capital goods freely available to domestic industry, he said. Kanwar, who maintained that the domestic industry had been over-protected all these years, however said that should continue to enjoy protection though only to minimal levels.

Two captains of industry who preferred to remain anonymous, however, said that domestic industry which was already high cost and inefficient would suffer substantially with the opening of the flood gates to imports. The high cost of industry was essentially due to the higher administered prices, which in turn were the result of inefficiency in the public sector.

The Scales Turned

Until very recently, the World Bank and the IMF generally applauded India's economic achievements and gave her good credit rating. Suddenly, in about a year, the scales turned against her and the rating dropped down the bottom exposing the underlying ugly and dangerous situation. Economists have agreed that this crisis situation was not the result of the Gulf war but the consequence of the policies pursued during 80s. Indeed, one can trace the crisis to the very structure of planning and political economy that were imposed on India by the Nehru-Mahalonobis model and its bureaucratic implements. If India has found herself suddenly faced with a large double deficit, budgetary and balance-of-payments, which threatens to bring about the collapse of the economy, it was the burst out of trends specifically set through faulty liberalisation during the eighties. During the same period, the growth rate of the economy went up by one and half percentage point. The economists were jibilent about it until it was found that the aforementioned double deficit sustained the growth which now is returning to its historical trend.

Over the last few years, the World Bank had exhorted India to reinforce policies of liberalisation that started in the early eighties in order to achieve higher growth rate, increase commercial borrowings, relax import and export restrictions for technological upgradation and competitiveness, loosen Government controls and put greater emphasis on exports. All this was recommended amidst a growing balance of payments deficit and crisis. At first, the reason given for the crisis was that India had "fallen a victim to its very success". Later, the blame was first put on the prevalance of controls and mismanagement of the economy. How those suggestions were related to one another was never explained.

It will be shown that so long as the focus is kept on regulation vs liberalisation keeping all structural issues untackled the crisis can only deepen further. The reasons are not far to seek. They are moving in a vicious circle. During the eighties, it was found necessary to undo the stagnation in industrial growth for which a number of steps were taken. An important one was to liberalise inputs of raw materials and capital goods. This led to stimulation of those industries, particularly consumers goods industries, which in expansion went for high import intensity. The other side of this policy was to award import replenishment titles to foreign exchange for export promotion which, in turn encouraged import intensities of the exportables. No wonder, while the growth rate of both industrial production and export increased, the trade gap kept on widening throughout the eighties. There was a demand for a further dose of liberalization without demanding any shift in priorities towards less import-intensive and more labour intensive industries.

This sequence of policies and development was reinforced by still higher fiscal deficits and Government consumption expenditure for ensuring a high level of domestic demand. Consequently, a vicious circle which began with limited liberalisation caught hold of the economy. A new push towards industrialisation without structural change accompanying it led to rapid increase in imports that widened the external deficit. Attempts to increase exports led to increase in capital and raw material intensity of industrial growth, imports as well as exports. Growth of employment in industry and agriculture declined. Thus, every negative development reinforced the other simply because liberalisation and industrialisation were left structurally rigid.

Consequently, the old relation between external and internal deficit was disturbed if not reversed. Earlier, the two deficits cancelled each other's impact on prices. Now these two deficits along with the implicit devaluation of the rupee created a large inflationary potential in the economy.

Liberalisation, like regulation by itself, is neither good nor bad; it depends on what it fulfils and what it negates. Many people still do not appreciate that the regime of controls and regulations of the first three decades put us neither on the path to rapid industrialisation nor on self-reliance in any significant way. On the contrary, after initial push, it induced stagnation in the growth of the economy along with new dependency and massive corruption. Under the so-called liberal regime of the 80s, the growth rate went up by one percentage point but it produced four disastrous results of a) debt trap, b) more rampant corruption, c) technological dependency, and d) environmental degradation. There was utter confusion. Although, there is a general consensus in favour of liberalisation no one has yet produced the right package of policies for it.

There are four important conditions or bases for liberalisation without which the economy can move into a double crisis of even deeper domestic stagnation and external dependency. First, import substitution must be accepted as inseparable from export promotion. Indeed, the two must never be separated as the experience of successfull economies like Japan, Korea and some others tell us. So far a curious aspect of Indian import substitution has been its degeneration into export substitution. The critical question is that of the sequence. Both the Japanese and the Koreans developed the technique of switching from one to the other and back as the situation demanded. Neither can be treated as a sacred cow.

Second, external liberalisation without domestic liberalisation may so induce and accelerate growth in one sector at the cost of others that the total outcome may be a continuation of stagnation and dependence. There is a powerful link between the external liberalisation and domestic controls and is reinforced by massive corruption. Nothing in the world can ensure the progress of Indian economy, whether regulated or de-regulated, if it cannot be rid of corruption. True, liberalisation by itself may help to remove some of the networks of corruption but not quite, unless those who run the state conform to its laws.

Third and probably the most important is the adoption of the right kind of technological policy. In relation to the now expanded structure, technological dependence is ever greater. The recent balance of payments crisis and credit sequeeze has dug out this ugly fact. A 1985 Reserve Bank study had shown that the regime of collaborations and controls, despite all the loud promises of technological self-reliance, made the economy more dependent on external technologies without creating a sound domestic base which would discard dependence whenever necessary. In the period of liberalisation during the 80s, technological dependence became even more pronounced. This is not an argument against liberalisation. It is an argument to treat liberalisation as a structural issue, to break the vicious circle.

Fourth, during the eighties there was a marked shift in the growth of manufacture of consumer durables as against non-durable goods. Not only is the former more capital-intensive and import-intensive, it also lacks international competitiveness so necessary for pushing up exports.

In the face of the current terrifying crisis and import intensity of industries, varying from 20 to 50 per cent, it is uniformally stressed that exports must be doubled at least every five years as a counter measure against larger external borrowings and repayment obligations and also to meet import obligations. However, in reality, the export performance though not unimpressive in recent years fell short of the objective of meeting necessary import costs and severe foreign debt. Indeed, the dependence on external finance increased even further. But if the level of import intensity is not reduced, and if it is increased under external pressures, the external deficit will widen. Global trends and policies of developed countries do not favour Indian exports. The crisis can be averted only by rearranging investment policies and import-export intensities.

The World Bank and IMF had always accused India of bias against exports and, therefore, strongly suggested dismantling of some of the export restrictions. But this time the purpose is quite different. Ultimately, when the cat was out of bag, what the World Bank wanted India to do was not only to liberalise imports but that in the absence of larger export, the country should resort to increased commercial borrowings. In the middle of the 80s, the World Bank recommended that India must borrow 18 to 20 billion dollars over five years, if it wanted to achieve a reasonable growth rate. This is what India precisely tried to do though not quite so successfully. This was ostensibly for achieving higher growth rate but the fiscal structure was such that it ultimately increased the external obligations leading almost to a debt trap.

Liberalisers Without Liberalisation 99

If all this and liberalisation too were not accepted, the country would have to settle for a lower growth rate. New Delhi, in panic, surrendered—in anticipation of a demand for surrender—by publicly annoucing that India need not be pressurised as she was introducing the suggested reforms on her own.

A Judicious Mix

Common sense demands that a country should employ a judicious mix of borrowing instruments, obtain both fixed and floating rate loans and make extensive use of suppliers, buyers and export credits for use in productive activities. Export was one such actively. These commonsense policies needed a determined effort to expand exports, if external economic relations were not to deteriorate further.

It is forgotten that a large amount of financial inflows that come in the form of aid are designed to provide budgetary support rather than making any real contribution to development. Aid collected for road building, schools, health centres and services should normally do not require any external input as all these programmes can be locally financed with rupee finance and resource mobilisation strategies. But the Indian Government has never stopped at any point and its representatives have been pressing on every country to give aid. It is a matter of great shame that India year after year is asking for aid from 29 countries, including such small countries as Belgium, Austria, Ireland, Norway and Denmark for purposes which deepen our dependence. In fact, the external component of Indian plan expenditure has been moving up and much more sharply in the industrial sector.

Occasionally, demands are being made for a big devaluation. Continuous implicit or explicit depriciation of the rupee, was thought to be a less painful method for adjustment then straightforward one-shot large devaluation. But it seems that the path chosen paved the way for the balance-of-payments crisis. There is no clear relation between nominal and real exchange rates. A continuous depreciation did not change the production ratio between traded and non-traded goods and hence, did not remove domestic constraints on exports. Most significantly, as was feared, it created expectations of future depreciation which acted as a powerful pull for dependency on imports and unnecessary inventory build-up.

Although, a large number of measures were taken to liberalise the economy externally, particularly easing of restrictions on import of capital goods and materials, and concessions were given to foreign companies in respect of equity participation, licensing and improvement of procedures, these external liberalisations were not accompanied by domestic liberalisation. Domestic controls on domestic economy were maintained, rupturing production for exports and domestic production.

If the World Bank package has to be accepted and implemented to achieve better results, it is absolutely necessary to have a corresponding domestic package in order to negate its undesirable consequences and to optimise objectives. The package must be shrewdly put together as to resist the damaging demands of the Bank-IMF as well as be complementary to measures of external liberalisation. This was not done in the past because the Indian bureaucracy could not allow the economy to be freed while it enjoyed some fruits from external liberalisation.

The real danger is that foreign trade strategies may become so demanding that they may overtake the Eighth Plan by default. The planners have a duty to restructure the domestic economy and to be deflected from the objectives of employment, growth and social justice. We are in the last decade of a century, away from the position where trade for the sake of trade was a national priority as in the case of Japan and Germany. These nations first used trade for domestic enrichment and equity and then for grabbing the markets of the world.

CHAPTER 11

Some Structural Issues

There is so much talk about hard decisions that there is a danger of our going back to bad old days of the late fifties and mid-seventies that wreckled the economy. Without massive and long-postponed structural changes, the thundering wheels of disaster cannot be stopped. Indeed India is passing through a series of economic crises which, if not met headlong with drastic change in policies, will surely take her towards a precipitous collapse.

Balance of payments and fiscal deficits have become so large as to expose a near bankruptcy. But these deficits are not *sui genris*. They are expressions of manifold structural crises and underlying mismanagement and corruption. It is possible to be so obsessed with deficits that one may let the internal crisis deepen. But one cannot continuously borrow or default on external payment. In panic, the Government is looking for new external loans and credits. What is not realised is that both the balance of payment and fiscal deficits are the outcome of the Nehru-Mahalonobis model, which has over the years created structural distortions and yet its drum-beaters remain unwilling to give up even if it means surrender to IMF-Bank blusters.

A structural problem is the dislocation of the relations between two or more components constituting a structure. These dislocations can always be corrected in a non-pathological structure. But if they refuse to be corrected it means that when the choice is being made between alternative policies, both choices are wrong.

The structure itself requires dismantling or drastic transformation. Indian economy and the ruling class and the elite are caught in this pathology. Quick remedies being suggested are not real solutions; they are part of the problem.

Inflation without end, poverty without cure, unemployment and underemployment structurally built-in, shifts within GDP towards less productive sectors, an unprecedented wastage of resources, a decline in the rate of growth and efficiency of infrastructural facilities and above all a gross mismanagement, all creating a desperate situation for vast masses. This crisis has been brewing and deepening for quite some years now. Within its chosen framework and development strategy, the ruling elite has come up against structural bottlenecks that remained but now are creating social and political conflicts and instability on a rather wide canvas. Indeed, the political and economic crises have been so interwined and enmeshed that their mutual reinforcement blinds us to the primacy of the political crises and the threats of international dependency.

Although the economic crisis is the direct product of the political crisis, some aspects of the crisis originate from the failures of the Nehru-Mahalonobis model itself, though indirectly they are related to the political crisis. The model made no specific political assumption except that Nehru was the political god that, of course, ultimately failed.

In a nutshell, the economic crisis is attributed to (a) structural problems or rigidities, (b) strategy of development itself, (c) policy defects, (d) growth of economic parasitism, (e) massive and unproductive controls and regulation. About the last, a business house head, Aditya Birla, remarked the other day: "I always tell my foreign friends that if a corporation can do business in India, it will be successful anywhere in the world. If an entrepreneur, for instance, wants to start a new industry, it is like entering into any obstacle race". To this obstacle race contributions are made by the aforementioned factors, leading to delay, increasing the gestation lag, industrial sickness, pushing up the cost and the capital/output ratio and finally creating a new kind of vicious circle for maintaining even modest growth rate and massive exploitation.

Difficult to Explain

It is rather difficult to explain how (a) an economy which has gradually arrived at a rate of saving of 22 per cent should get permanently stuck with less than four per cent average growth rate, (b) the farther the development proceeded, the more people, proportionately and absolutely, got pushed below the poverty line, (c) eight States which are rated as resource-poor have got more developed and the remaining, most of which are resource rich have stuck around a zero per capita growth rate for more than ten years, (d) the more the investment, the less the return on it (e) the commanding heights of the economy contributed to the sinking of the economy, (f) even in that part of the agriculture for which every input was available, including fertile land, yield per acre stagnated, (g) with cheapest labour in the world, we have developed the most cost-bloated economy; (h) with every Government committed, largely for its own survival, to a public distribution system, the distribution remains elusive and most difficult problem to tackle.

One does not need the support of any deterministic theory to prove that the crisis of the Indian society is basically both economic and political. And this crisis has been brewing since the mid-sixties when the Congress party lost its hold on the polity. Mrs. Gandhi's populism, to which I was the first to draw attention in 1969, let loose a hell on the polity that ended in Emergency with the support of the drum-beaters of the Nehru-Mahalanobis model. Populism was the inevitable product of the model though it also sounded its death knell.

If we take 1965 or average of three years around that year as the dividing line, the whole of the post-Independence era can be divided into two parts. This divide shows that the policies persued in the first period led to a general declaration in growth in the second period and exposed the Nehru-Mahalanobis as a model of famine, poverty and income inequalities although along with a modest rather than high industrial growth. If we further divide the second period into two with the eighties falling in the latter we find that from 1965 to 1979 the growth rate of the economy declined but went up again to 5.2 per cent in the eighties but at enormous costs.

But the cost factors proved alarming. Leaving out the incalculable ecological costs, the external debt increased by five times and internal debt seven times. The rate of inflation has been twice that in the first period. Reliable statistics are not available, but the concentration ratio and the Gini coefficient remaining the same it is clear that development had no impact on distribution. This consistency, however, is doubted by everyone because even the NSS itself has no faith in the expenditure revealed by the upper two deciles of the households. Since the data do not take note of black money, statistics about the pattern of distribution if properly adjusted will only show a deterioration of the economic life of the bottom half of population. There was no reason for this half to support the other half, particularly, the minority ruling elite. A revolt was brewing and has now come on the surface violently. Now, in view of the new trend of static or declining employment elasticity both in agriculture and industry with a modest per capita income growth, only a very naive person can imagine a perceptible dent on poverty.

Despite the enormity of the economic crisis, structural bottlenecks and near financial bankruptcy, the Indian economists seem to be still engaged in the merry traditional analysis of ups and downs of growth rate which fails to capture the magnitude of the crisis and, even more tragically, fine-tunes the same policies which created the crisis in the first instance. Although no one is listening, the economists are debating furiously among themselves on rusty issues.

The transition from one socio-economic structure to another and the simultaneous realisation of a high growth rate have had to be made consistent. The change in the socio-economic structure would require making contradictory aims compatible with each other by giving them some socially determined numerical weights. In fact, it is technically feasible to make all priorities and objectives comparable if only some cost-benefit criteria can be formulated. But, in this way, the contradiction is not removed; it is only camouflaged by the relative values assigned to each one of them. However, one important reason for planning is that the composition

Some Structural Issues

of the planned increment to the national income has to be predetermined in terms of a given product mix. But one product mix gives a different level of GNP from that given by another. If the objective is to maximise the GNP as a social measure of well-being, then the question of product mix is not subject to any social criterion. It means applying the principle of economic rationality and cost benefit analysis at both the macro and micro levels. The poor masses do not seem to exist; only products do. Karl Marx failed to imagine a situation of commodity fetchism with declining share of commodities in the GDP.

The main transformation problem set was to convert a poor, static, predominantly traditional agricultural economy to a richer, dynamic and modern industrial society in a reasonable period of time but without any micro planning. It is no accident that every public sector unit ignored micro planning and every private sector unit was so constrained by controls that no rational micro planning was possible. Every decision was subject to bribery and corruption. Yet, the overall structural economic transformation was calculated in terms of changes in output composition as, say, between primary, industrial, service and utilities; distribution of labour force in these sectors; changes in demand, imports and exports and substitution of domestic production for imported goods; incomes, rights and opportunities to be determined by skill than by tradition. The scale and composition of this transformation largely determine the size, pattern and quality of economic growth. The share of the poor in the national cake was left indeterminate. The whole economic game is confined to how the non-poor divide the national income among themselves.

Indian planning included, in theory, all the known strategies but in practice these wre restricted to four: the growth rate, public sector, some capital goods and resource mobilization. The last was required to effectively sustain the plan, but because of the failure, erosion or downgrading of the other three strategies, the strategy of resource mobilisation also proved self-defeating. For example, the proportion of tax revenue to national income increased continuously, but because of massive evasion, that increase was sustained only through a system of steep graduation, which, in turn, encouraged further evasion so much so that ultimately a black economy emerged as a parallel economy and ultimately the tax proportion refused too increase. Indeed, it declined. Institutional savings, which increased along with increase in tax revenue, showed first a relative and then an absolute decline, thus divorcing the mobilisation process from problems of structured transformation.

A Paradox

A structural-cum-ideological paradox of Indian planning has been that the Nehru-Mahalonobis model accepted the amazing philosophy that a capital accumulator must become poorer by accumulation and that if the public sector got fattened by de-accumulation i.e. capital consumption it was bad but it needed no reversal of policies. Before even the economy began to grow, it was constrained by ideological considerations. The public sector failed the nation and sponged on the savings of the households from year after year with mounting losses. Ironically, the only policy constraint kept on the rate of growth of the rich or concentration of economic power was to restrict the rate of accumulation in the private sector. The stated assumption was that in order to reduce the concentration of capital and wealth in the hands of the private sector, its profits must not be high. High taxation should take care of it. The unstated assumption remained concealed. It was that if profits declined, the rate of accumulation in general must also decline. This was asking for suicide. The gap was to be made up by the public sector savings. But the sector became so inefficient and corrupt and loaded with so-called social obligations that the larger the investment, the smaller was the surplus until the surplus disappeared. However, the investment momentum was kept up by foreign aid, budget deficits and mobilising bank deposits. The last one required encouraging capital accumulation of the households, meaning the middle or the upper classes. The public sector appropriated these savings through borrowings which further fattened these classes.

If someone has capital for investment, he must be expected to become richer at the end of his investment and not poorer as a

Some Structural Issues

result of the outcome of his investment. The return on investment had to be reasonable. The rich have to become richer as under any private system of totally free enterprise or of any mixed economy model. The only structural corrective was to make public sector earn bigger and bigger surpluses and then expand it at the cost of the private sector. But this was not to be. The fiscal correctives were carried too far and became counterproductive. Mrs. Gandhi did not care about the economic consequences of marginal rate of tax of 97.5 per cent, introduced with the support of Marxists, Neo-classical economists and Keynesians as well as the bureaucrats, as that policy opened up new vistas of conspiracy to defraud the public exchequer. Corruption became the factor of last resort to bring about the equillibrium which the economists are so very fond of. Without a new consensus of property relations, we will remain trapped in the old sterile debate, the terms of which were set by pseudo-Marxists and cynical Nehruvians.

The Nehru-Mahalonobis model, whatever its strong points and defects, was structurally a supply side model insisting upon public investment for the production of goods for which there was not yet adequate demand. The alternative supply side assumption was that public sector investments could create such supply conditions for which adequate demand would be automatically generated. This provided a correct assessment for some sectors such as power and coal but not for others. The results are there for all to see as shown in the public sector with large unutilized capacities, eating its own capital and the private sector usurping its use with the support of the public sector itself. In general, the business community complained perpetually of demand recession. It is not often noted that in more years than less the Indian industries faced stagflation. On the other hand, demand management through fiscal measures which failed to reduce inequalities led to massive corruption and ultimately collapsed.

Ironically, this supply side economics had the powerful support not only of the Left but also of the bureaucracy, a curious mixture. It naturally began with high priority being given to capital-intensive goods sector. This sector was, by definition costly both in private and social cost because of a high capital-labour ratio and was designed to push production towards consumer durables for which ironically there was a demand as against goods of mass consumption for which the demand was weak on account of poverty and a very slow rate of employment growth. At a psychological level, this priority satisfied the nationalist sentiments and romantic ideologists who wanted India to industrialise fast but without tears.

The neglect of demand management was the basic defect of the Nehru-Mahalonobis model. Too many controls and regulations created a variety of demand-supply imbalances and their repeated suppressions produced many structural bottlenecks. From the demand side, there were two main factors that complicated and structural problems. First, pressure on the labour market depressed labour income on account of the rate of employment generation being far below the rate of people entering the labour market. Second, the bias towards capital intensity also produced a bias in favour of general material intensity which relied on heavy imports undermining the balance of payments deficits.

The most disturbing aspect of the structural stagnation in the economy is that poverty, despite all the noises made to remove it, has come to be accepted not only as necessary for maintaining the living standards of the top 10 to 15 per cent but implicitly encouraged as a structural barrier against both inflation and revolution. Indian inflation in contrast to other third world nations has been quite modest. It has been so because the employment and income of the poor are such, as to forcibly maintain half the population on or below the subsistence level. Another 20 to 30 per cent are only marginally better. It is impossible to expect these people to revolt against their masters as their immiseration has sapped their capacities.

Notwithstanding insistance by the Government and the Planning Commission on the decline in the percentage of persons living below the poverty line, a claim which no non-official economist is prepared to corroborate, it seems that the poverty debate and the anti-poverty programmes of the Government have proved ineffective in making a real dent into the problem. The level and depth of poverty, more or less, remain where it was when the

Some Structural Issues

debate started. The pathological nature of the problem was sharply exposed by the fact that the more anti-poverty programmes were mounted the bigger were the holes for leakages and less the concern for the problem. The new policies maintained the political opportunism of the old by recognising the failure of the policies and even highlighting them through the officially controlled media. The end seemed to ensure a high degree of public de-sensitisation. The fact that the poverty debate has nearly ended without ending the poverty is a clear expression of this de-sensitisation.

But the wheel has turned in a rather vicious way. Instead of radical social change, we witness widespread individual and group, often purposeless, violence, matched or even provoked by state violence. Political assasinations are the order to the day. The order change is the acceleration of inflation as a result of mounting budget deficits, increasing wastefulness of Government expenditure, rise in administered prices and corruption. A normal difficulty has become a structural pathology.

A Dangerous Situation

In fact, these contradictions have created a dangerous situation: a combination of repressive forces at the lower layers of society where poverty exists and the economic and political power at the higher echelons where economic power rests, leaving a dangerous vacuum in the middle. Another serious contradiction has been populist rhetoric about poverty on the one hand and decline in the growth rate on the other. The resources were diverted from the investment pool to the so called anti-poverty programme but the poor never got them and the economy stagnated or ran into crises, a case of structural contradiction.

Unfortunately, the emphasis still remains on poverty alleviation and not elimination. It would be the cheapest programme in the world if by spending about a few thousand crores of rupees, 25 million families i.e. about 120 million people could have been brought above the poverty line as indeed was claimed by the Rajiv Government for the Seventh Plan.

Poverty alleviation programmes failed because the state policies relied almost totally on subsidies which were legally or illegally appropriated by the middle classes, both in urban and rural areas. These subsidies which now take away about 45 per cent of both the Central and State budgets, are now causing fiscal bankruptcy but without having benefitted the poor.

A critical structural problem is one of disruption that has occured in the circular relations between pattern of investment, production, saving and distribution. It is well known that a very high proportion of assets and capital, about two-thirds according to official statistics and three-fourths if black welath is taken into account, are in the hands of the top 20 per cent of the population. As mentioned earlier, it is this class which both saves and exhibits a high level of demand. Surely, if they consume more they should save less. But they do not. How can they do nearly all the savings and all the consumption relevant to industry? It is due to the missing link in the aforementioned circle, namely black money and employment. Without employment becoming an explicit variable on the circle, disruption is inevitable. It is amazing how Government and the Planning Commission thought that they could reduce concentration of wealth without a full-employment strategy.

Only in a society where all means of production are in the hands of the state can the concentration of wealth be avoided. It is no accident that even the most advanced capitalist countries where welfare state has been used as a countervailing mediating force to correct the ravages of capitalism, inequalities have increased but not declined. However, in their case the most significant achievement has been that the income at the floor has been constantly increased along with increase in growth rate by putting full-employment in the centre of the strategy of development. Indian planners never developed such a model. Even before the growth-strategy was launched, capital deconcentration was put high on the agenda. Growth was stunted right at its birth. The structural link between capital accumulation and employment still remains broken.

The sterile and phoney debate with regard to public versus private sector is still being continued as if we are living in the fifties. The net outcome of policies of regulations, controls, licencing and price fixation has been the generation of the largest rentier class in a world. This class fattened for no other reason except that it had control over inventories and foreign exchange. No Indian economist, radical or liberal, has been able to integrate this class into the development models of class relations, income and welath distribution and determination of factor incomes.

If the structural bottlenecks in production have to be removed and also a high rate of investment to be maintained, then both the public sector and the private sector must be treated as one sector for creating new surpluses. Only that level of taxation should be imposed which would not be considered prohibitive for purposes of accumulation and saving and for further investment and also not give an incentive to economic activity being pushed into the black market economy. Only that policy of regulation should be permitted which leaves no discriminatory power with the bureaucracy. Attempts to solve the structural problem by expanding the public sector, regulation and promoting a welfare state (as against one for human resource development) have collapsed everywhere and more so in India. A hundred eighty degree turn towards the market economy may seem tempting but it cannot solve any structural problem.

Among the structural bottlenecks, the most important has been the low growth of employment not only in relation to growth of population and labour force but also in relation to choice of technology of capital intensive as against labour intensive choices. Although the average growth rate in the last four decades has remained modest and subject to serious fluctuations, its increase in the eighties by one-half percentage points was accompanied by a fall in the employment elasticity in both agriculture and industry. But the debate once again is getting zusty of all the structural changes required the most fundamental is to shift growth to employment. Most probably, we will get a higher growth as a result of this shift but it would be a different kind of growth. For that to happen, not only should the Nehru-Mahalonobis be thrown out but a structural change in the political power also be brought about. Who will bell the cat? We are seeing the rise of peasant power.

The relations between agriculture and industry has assumed new structural dimensions. The share of agriculture in the GNP has declined from 60 per cent to about 35 per cent in 40 years but the porportion of people depending on it is more or less the same. This dependance is aggravated by the stagnation in rural industrialisation. Indeed, the ratio of urban to rural population has not significantly changed in one century and under the prevailing strategies is not likely to change. Neither the growth rate can go up beyond 5 per cent at the most nor unemployment has much chance to decline. The growth will remain stagnant because of the demographic pressures.

In contrast to that, the share of manufacture in the GNP has also become static hovering around 22 per cent, thus showing that a classical industrialisation was an obsession of the Nehruvians. They failed even to explain what had happened. No wonder, Deepak Nayyar (1978) explained the deceleration of industrial growth in the second period in terms of income inequalities and the consequent limitation on the expansion of markets of mass consumption, whereas Pranob Bardan using the same NSS data refuted Nayyar and looked for an explanation in decline in the growth of capital goods industries. The late Sukhamoy Chakravarty explained it by the decline in investment in the public sector. This, in turn, was the result of budgetary deficits. At this point, the buck stopped. The Nehruvians could not attack the public sector as it was a sacred cow, even though instead of giving milk, it was drinking it. The public sector had either to go or to prove its worth.

The Main Stay

Paradoxically the increase in the share of tertiary sector, which is largely unproductive, has become the main stay for demand for manufacturing goods. The tertiary sector is taking away investment resources, a trend which ultimately will have a negative impact on growth itself despite its low capital-output ratio. A decline in the share of commodity production will block long-term growth. But the paradox is that if in the short term the tertiary sector declines, so will the manufacturing sector insofar as the former creates large demand. The economy is caught in both a cyclical and structural strait-jacket on account of peculiar relations between agriculture and industry.

Not only did the Nehru-Mahalonobis model neglect agriculture—as if it was no part of the development—even after severe agricultural constraints appeared, but the false old argument was also advanced that the Indian industrial growth had a largely autonomous character as flows between agriculture and industry were not sufficiently strong. In 1987, when the crops failed but industrial growth was kept up it was cited as the final evidence of the clinching of the arguement. In reality the rural-urban dichotomy that was produced by the Mahalonobis model lay at the root of structural disruption and rupture in the economy. More seriously, Indian industry has been increasingly made dependent on the external world which not only increased its *compradorization* but brought about a divorce between industry and agriculture as to depress both.

Another structural crisis is that the larger the investment made in the infrastructural sector, not only the shortages and losses incurred the greater but they have also now become structurally embedded in that sector. Power, transport, communication, energy, all of which are interconnected through backward and forward linkages, are short in supply notwithstanding the fact that they take more than two thirds of the official plan's investment resources. Indeed, this sector is now planned to run on deficits. This need not have baffled the economists but for fear of being dubbed as reactionaries they have turned a blind eye. A sector whose planned expansion is based on deficit is nothing but either a pathology of policy or a structural straitjacket. It is even more amazing that the Government deliberately provides inadequate investment in this sector particularly power. Deficits and shortages are further accentuated by mismanagement, corruption, lack of accountability and political and civilian interference into the business enterprises.

Talking of mismanagement, it is no longer purely an economic or administrative issue. It is a question of total political culture as well as of institutions. Indian democracy is an institutional arrangement but its institutions have been perverted, eroded and immobilised by the excessive corrupting of power by the power elite, be it business, bureaucracy or the politician. India has gone through a long institutional change the kind not witnessed so far in any Third World country except China. But, at the same time, institutional destruction at the hands of the populist regime has left unresolved the structural issue of Indian polity.

Institutional destruction was partly brought about by the bureaucracy which aggrandised its power through controls and partly by the prolonged phase of Indira-Rajiv power psychosis. It created a situation of conflict and corruption.

The artificially created conflict between the business community and the political elite ended up in the creation of the most powerful mediating instrument, namely, the bureaucracy, which became the creator, and ultimately the biggest beneficiary of the regulatory, control and licensing system. The bureaucracy performed three functions in this game (a) it both formulated and implemented regulation and control policies and kept out every other element from this power; (b) provided the link between the business community and the politicians; and (c) became a class by itself through excessive proliferation and assumption of an autonomous role. In the Marxist terminology, the Indian bureaucracy ceased to be strata long ago; it has become not only a class in itself but also a class for itself. Without changing the relationship from one of conflict to cooperation to which the dismantling of a large part of bureaucracy is a pre-condition, no structural change is possible. Bureaucracy is the worst enemy of institution.

It is not possible to enumerate the structural terms and dislocations. However, what has been stated above provides enough ground for pessimism, if only because the political system and the power elite would not permit adequate structural response and institutional autonomy.

CHAPTER 12

Siege by Growth Fetishists

The post-war experiences of successes and failures of the former colonies suggest two conclusions. On the one hand, a high growth rate around 8 to 10 per cent if persistantly maintained, irrespective of any other aspect of economic life, gives a success story in terms of employment, standard of living and even distribution. On the other hand, a modest growth rate of about 3 to 4 per cent, again kept persistantly at that level, is a failure. Any comparison with the advanced countries or with pre-colonial situation will be entirely misleading.

A modest growth rate need not be a disaster if its benefits are fairly distributed, guarantees full employment and ensures a zero population growth rate. In the absence of these conditions, obsession with the growth rate, when the range of change is structurally or functionally rigid, can turn into a pathology and lay an intellectual and material siege around the society. India falls in the second category. All the models and planning strategies opted so far had bias against employment generation. So much so that the last decade saw an accelerated growth rate but a decline in the growth rate of employment. The population growth rate did not register even a slight decline. No wonder, after four decades of so-called planned development, almost all the conceivable crises are converging on her.. It is my argument that without replacing the economic growth rate by one of employment growth as the predominant strategy, India will remain a failed case. Those who cling to growth rate are the siege makers.

Attack from the Vested Interests

Ever since Mrs. Indira Gandhi threw out Prof. D.R. Gadgil from the Planning Commission, the Commission's authority and legitimacy continued to be eroded by political and bureaucratic interference. Mrs. Gandhi did not understand anything of economics, let alone planning. She used the Planning Commission to push policies arbitrarily towards political manipulation as it suited her. The rot started by her continued until Mr. Rajiv Gandhi declared that planners were jokers.

Paradoxically, India and many Less-Developed Countries (LDCs) while having achieved a high rate of savings and capital accumulation and near sufficiency of wage goods at existing levels of income and distribution, have neither fulfilled the targets of high growth rate nor were able to remain on a steady growth path of the targeted rate range and certainly did not ensure social justice. Indeed, problems of poverty, population, growth inequalities, malnutrition, and illiteracy have become serious and complicated. On top of all, despite a high rate of savings, the two gaps—the budgetary and trade—have widened instead of getting narrowed.

Yet, India continued to plan on the principle of more of the same in the hope that things would improve in the long run. Our continued faith in the borrowed growth models that failed, even when the long-term perspective seemed to recede, was a clear expression of an open society with closed minds. Indeed, there was no long-run except a series of short-runs. If structural imbalances, of which there were quite a few, were accentuated from one short-run to the next, the crises were bound to deepen. The ruling and privileged classes all seem to have vested interests in perpetuating the old development strategy that revolved around the growth rate. As time passed, the increasing emphasis on the growth rate, ironically, concealed the challenges of underdevelopment for which growth was enthroned in the first place.

The growth rate was higher in the eighties over that in the previous 15 years which in itself was lower than what was achieved until the mid-sixties. Nevertheless, the average 4 per cent for four decades is too little against all the tall promises of the growth planners and for the trickle down economists. A 5.3 per cent average growth achieved during the eighties brought the *growth wallas* back into the game again even though it was achieved at a prohibitive cost and with total fiscal anarchy.

Besides, there had been a drastic shift in the structure of the GDP from *productive to relatively less productive sectors,* from primary and manufacturing to tertiary sectors whereas the percentage of population in the rural areas has only marginally declined. Our preoccupation with sectoral growth rates, investment allocations on the basis of sectoral growth requirement, our assumptions about growth trickling down, our faith in macro variables and averages concentrated too much on wrong directions, priorities and policies. One does not have to produce more statistics to condemn the trikle down theory. It stands condemned by the logic of the growth rate remaining at the centre of the plan.

Vested Interests

There are several vested interests against a development strategy that is not centred on the growth rate and focusses instead on growth of employment. At least six such interests can be identified. The first represents those who believe that the mainstay of the Indian economy is the market for 100 to 150 million people, the middle and upper classes. For this class, apart from the infrastructure necessary, no matter what strategy is adopted, the focus is mainly on the development of capital-intensive and high-quality consumption goods, particularly consumer durables. Ironically, it is this class which is responsible for high household savings as much as for vulgar consumerism. From Adam Smith to Karl Marx to J.M. Keynes, this class defied all their theories. Yet, the basic philosophy of planning all through has been that as this class expanded the economy would make progress. The rest of the population had to survive at a subsistence level or below it. The growth of income of middle classes is considered the linchpin of the growth strategy on the one hand and internationalisation of the Indian economy, with or without liberalisation, on the other.

The second is represented by those who have tasted tremendous benefits and profits from the liberalisation policies of the eighties. The liberalisation policy was without any perspective and it was carried haphazardly and not often without kickbacks. All this was done in the name of building the international competitiveness of Indian industries and services. This was a patently absurd proposition because there are several sectors of the economy which have nothing to do with international competitiveness. But, no clear demarcation was suggested. What was needed was the insulation of small and medium industries from the pressures of external liberalisation and the encouragement of domestic liberalisation and competition. Without successful domestic competition, there can be no international competition of any meaning.

Third, the big Indian business which enjoyed sheltered markets for a long time at the cost of the medium and small sectors also now finds that through accelerated foreign collaboration they can profit from liberalisation without any new commitments to self-reliance such as R&D and exports. The big export houses have really not fulfilled their promises for export but they have pocketed all the import advantages, fiscal and otherwise. It is quite logical for them to believe that by pushing up the case for a high growth rate and international competitiveness without structural changes, their share of national financial resources would increase irrespective of what happened to the rest of the economy.

There is a strong lobby of the international financial institutions which insists on the integration of the Indian economy with international system and this they believe can be better promoted by creating a false illusion that only if the planners were to aim at a higher growth rate, the country would get more resources for private foreign investments. However, if one looks at the figures of foreign investment in India for the last few years, the picture is quite dismal. If we are unable to fight back Super 301, it is because both its supporters and opponents argue from the same assumptions about internationally recommended growth strategies.

Fourth, the set of economists and bureaucrats, who still are psychologically committed to Nehruvism, feel that their legitimacy will be eroded and their theories proved defunct if the direction of the economy is fundamentally changed.

Fifth, notwithstanding the failure of public sector and massive expansion of new developmental activities of the state that led to huge deficits, there is nothing to stop the proverbial ideology of centralisation keeping its stranglehold on policymakers. A major change that is needed is the federalisation of Indian planning which requires transfer of some of the economic functions, often misappropriated earlier, to the States. For example, a number of centrally-sponsored schemes have to be transferred to the States. The Union Ministries have consistently cajoled the Sates to accept them. With a higher growth rate, these ministries can come up with a proposal for higher allocations. More significantly, without the transfer of power from the Centre to the States, there would be strong resistance for further devolution from the States to the local bodies. Moreover, it does not occur to the opponents of decentralisation that without local resource mobilisation, high growth, let alone social justice and participatory democracy, is not possible.

Sixth, the senior bureaucrats who wielded tremendous power in the long years of Nehru-Gandhi era are not willing to shed the misappropriated power, particularly that of the Planning Commission.

It is my purpose to show from the experience of India and most LDCs that by enthroning growth rate, they ignored the need for structural changes in their economies particularly those relating to improving the quality of life, productivity and employment. Besides, the economic cost of growth in terms of ecological deterioration has been unduly high. Whereas growth has to remain an important variable, it is not to be the sole dependent as it has neither improved the quality of life nor ensured even its own objectives. Therefore, India and other LDCs have to shift their focus from growth to some other variable.

Models of Employment

According to Jorge Mendez "A solid body of theory on the subject of *employment and growth* in underdeveloped countries does not really exist".[1] According to him, there are four models of employment following respective theories of economics. First, the neoclassical theory states that full employment will be achieved if conditions of complete free competition exist because under free competition not only will the resources be most efficiently allocated, the distribution will also be such as will ensure rewards which will yield savings and investment that will be consistent with the capacity of the economic process to accumulate. The paradox that discredited the theory was its insistance that poverty will be eliminated if people are willing to accept wages below the poverty line. Wages will rise with increasing employment, accumulation and growth.

The second is Keynesian and the neo-Keynesian theory. There were two aspects of it: first to stimulate demand through public expenditure and second the investment in the productive public sector. The neo-Keynesian theory provided support for both, first for avoiding recession through demand generation, and the other through improving productivity and capacity created through investment in public sector. Demand deficiency caused unutilised capacity, surplus production and recession even at a rather moderate rate of growth. What was not clear was how the Keynesian policy by itself could generate full employment and also achieve equilibrium. Besides, in practice, Keynesianism produced a monstrous state, excessive dependance on state and distortions of markets.

The third view is attributed to Lewis, Fei and Ranis. This model is derived from the existence of vast surplus labour in the rural areas and is based on the process of transfer of this labour from agriculture to industry without raising wage rates. The wages will remain near the subsistence level until a very high level of employment is achieved. Agriculture will keep producing wage goods as the withdrawal of labour will not raise costs. Whereas Lewis insists on planning, Fei and Ranis suggest no interference. The only condition is the optimal absorption of labour through

---

[1] Yohanan Ramati, (ed.) *Economic Growth in Developing Countries 1975*, New York, p. 173.

production processes characterised by labour-intensive technology. How this can happen without planning distribution and choices in technology and priorities is not clear. Lewis has, as if, suggested a planning model for the theories of Fei and Ranis.

The fourth model that became very popular and widely debated for decades was put up by Paul Prebisch in his Transformation and Development (1970) in which he advocated a vigorous global policy with a high level of employment being given a pivotal role. His book is the culmination of the proposals that he had been putting through the Economic Commission for Latin America. Employment policy must be an integral part of planned development and related policies. But, the overall emphasis was still focussed on higher and higher growth rate because of what he called the need to remove the "dynamic insufficiency" of the undeveloped economy. Dynamism could be increased by a higher growth rate. He also spoke of "correcting the dynamic insufficiency of the economy with a great social orientation". This implied two things: (1) to provide the population with basic needs; and (2) to search for a type of growth that will create the maximum number of new job opportunities, both in industry and agriculture. The role of the rest of the world in providing aid and trade opportunities was again emphasised.

However, as Jorge Mendes concluded the research work such as that of Fei and Ranis, "helps to clarify the labour absorption process in the modern dynamic sector. But, both of them failed to appreciate the degree of absoptions that can be achieved in practice. They also failed because of their tendency to consider the development process as depending almost entirely on the capacity of the modern sector for growth, the other sectors remaining passive, depending on the opportunities that originate in the privileged sector". All the Prebisch conditions (1) easy supply of labour; (2) high growth rate of 6/7 per cent; (3) growth of modern sector; (4) large external resources are satisfied but the unemployment problem remains.

Critique of the Growth Theory

There are several well-argued and well-tested arguments for not giving central focus to growth rate. First, keeping focus on the growth rate shifts emphasis from processes, policies and linkages to a single index. When everything is pushed into a single index, a number of distortions take place, some of which may remain concealed. In a sense, this index covers up mistakes made in policies about processes, linkages and sectorial balances.

It is no accident that most of the theories of similar development revolved round the Harrod-Domar model. There has never been any escape from it, not because it was valid, but because it was a truism. The simplistic character of this model made development look more of a technical and mechnaical problem of internal balances than a matter of social and political engineering. For instance, the problem of accumulation came to be treated as the most critical, no matter how accumulation was achieved.

In 1946, E. Domar complained about the rate of growth as a "concept which has been little used in economic theory". Since then, the concept has been over-used, so much so that economists and other social scientists, other than econometricians, do not like to use the concept without adding to or substracting from it something.

Second, all growth models used three major variables, namely, capital accumulation, capital-output ratio, and population growth. It was the wisdom of the conventional economics. Besides, population control is not subject to policy decisions within an inequitous socio-economic system. Even more true is the fact that capital accumulation is not always subject to short-run policy decisions; only its rate can be modified. Total capital accumulation—not who accumulates it and at what cost—is the foundation on which the chief economic paradigm on modern economies are based. Unless this economic paradigm is replaced by another in which the capital accumulation is given sociological and distributive dimensions, one cannot give growth also a social meaning. Changing captial-output ratio is matter of both macro and micro policy decisions, but those decisions go far beyond the growth target.

The third conceptual problem emanating from growth relates to flow-stock relations. Economic growth is defined as a measure

Siege by Growth Fetishists

of 'flow' rather than 'stock'—the flow being the GNP. High flow growth is identified with good political management, not treated as an indicator of development and expression of welfare. For instance, we contrast India's low energy consumption per capita with others having higher per capita consumption. Yet, we use even this energy very inefficiently and often distort its end-uses. To disregard the concept of stock is a road to economic suicide as we are now beginning to realise. Energy, minerals, water, land and air are to be treated as both flow and stock and not free goods. Flows always depend on stocks. We need a new stock-flow perspective for making sense out of the growth rate.

Its corollary is that the net resources available must be clearly distinguished from the gross resources. Since depreciation has been increasing at a faster rate than net savings as it always does in early stages of industrialisation, the resources for the Five Year Plans have been artificially boosted. The cost overruns have become so scandalous that the distinction between net and gross has become of central importance. But, even the Planning Commission has not come to grips with this problem.

Fourth, the growth rate has encouraged mathematical model building or alternatively by their very methodology, mathematical model building and econometrics have brought the growth rate to the centre of the strategy and theory of economic development. Since all such model building is based on certain assumptions, things can go wrong if assumptions are changed. Leontief had to admit that "If you do not like my set of assumptions give me another and I would gladly make you another model". A new problem is then created, namely, the testing of assumptions. This provoked another economist to say, "when it comes to normative models, though frequent in literature there is even less obligation to test assumptions."

Fifth, the trickle down was considered a natural—even inevitable—economic process for generating continuous growth impulses. After the central failure, growth was made synonymous with development. Later on, technology and development were also treated as synonymous. Needs were confused with wants; dynamic relations between short-term and long-term development, sometimes negatively correlated, were missed. Most significantly, limits to growth were seen in terms of skilled labour, capital or technology. Above all, the global social cost of economic growth in terms of ecological and environmental degradation were either assumed or were considered to be problems which could be easily handled by trade and new technologies. No wonder, we now face every contradiction that was put into the cauldron of international trade with the expectation that expanding trade will take care of many problems; trade expansion will push growth, technological development and cost reduction through new revolution and communication. But the cost of trade is not fully taken into account in the growth, particularly when protection walls are being raised.

Following Kuznet's inverted U-curve, it was assumed that through the working economic forces, capital accumulation would bring a change over from capital accumulation to social welfare. The theory of turning points was evolved about this change. Even Arther Lewis, Singer and Prebisch were all working and hoping for turning points to turn up. The difference was that "early models of developing thinking were not heartless, or inhuman", but gave priority to the enlargement of the cake before it was to be distributed. Thus, the trickle down was explained more elaborately. It was capital accumulation through inequality; it was sequentially arranged; it led to the cake-first theory. If all this succeeds, welfare will increase. Of course, welfare did not increase, but the growth theory remained enthroned. It is globally well established that income and wealth inequalities have not declined significantly over the past half century, and probably have not declined at all in the past quarter century.

Sixth, linearity assumptions about growth, although challenged by some, have generally been accepted and proved the graveyard for many policies. Most economists, particularly non-Marxists, assert "that the growth of a properly functioning economy is linear, gradual and continuous. It proceeds along what economists have called the MIT Standard Equilibrium/Growth Curve... (they) see no necessary connection between the process of economic growth and political development such as war and imperialism;

these political evils affect and may be effected by economic activity, but they are caused or essentially caused by political and not by economic factors. Liberals believe economics is progressive and politics is regressive."

There are conflicts between linear quantitative growth and qualitative deterioration. For instance, in India, nowhere has the concept of growth rate done greater damage than in the field of education and educational planning. On the one hand, it has provided a narrow perspective of creating such a manpower as may be needed for the economy. On the other, expansion of education has been so anarchic and so divorced from well-known ideals and objectives that the system can be easily declared as an instrument of both internal and external cultural domination. India is one of the few countries in which the poor pay for the education of the upper classes. Besides, no country in the world makes education subject to compulsions of growth rate only and yet lets the rate remain subject to political pressures.

Seventh, a five-year growth target assumed a steady growth, even when allowing for small ups and downs. Growth rate, both in private and public sectors, is made a function of investment and is delinked from risk and accountability. In modern economies, risk and uncertainty are present all the time. Growth rate strategy assumes stable expectations. We have not yet learnt the lesson that average growth rate over five years is operationally meaningless if very wide fluctuations take place from year to year. An economy which still depends for growth on the vagaries of monsoon cannot remain on a steady path. Yet, the planners assume not only regularities in expenditure and investments, a continuous flow of funds from the savers, but a totally crisis-free economy, despite the experience to the contrary. Every Five Year Plan has suffered many crises and distortions. If it has survived it was because the cost of crises and distortions were excluded. For instance, the shortfall in food production during two successive droughts may be made up in the next two good years, but the costs and terms of deterioration of health of the people and death of animals remain incalculable.

Eighth, it has been amply demonstrated that growth and development are not co-terminous and sometimes they can move in opposite directions. Besides, though it is possible to aggregate the components of growth, such an exercise is not possible in respect of development. No matter how it is defined including the addition of quality of life indices it cannot be given a single focus. Development, even if definable, is a complex concept and cannot be measured. Even when it replaces growth, development cannot comprehend several objectives such as distribution, regional disparities and the rural-urban divide.

Nevertheless, it is agreed among economists that the concept of development has to replace growth, whatever the difficulties. For instance, it was suggested that a set of social welfare programmes be added to growth models in order to expand the area of planning and to meet social objectives. But this additionality could not correct the ravages of growth. Yet, this additionality along with growth was called development. In short, whereas it was recognised that focus should be on development, when it came to measurement, growth rate again became the main determinant.

'South Asian Child'

Based on several decades of experience, the UNICEF in its document entitled 'South Asian Child' has summarised its findings in the following words: "A relative neglect of social development continues to act as a drag on economic growth, despite the priority given to the latter. Social attention to the multiple need of a family or a group is fragmented and uncertain—in terms of food and nutrition, clothing and shelter, sanitation and health, education and employment. Community and even family support is uneven in relation to the specific needs of human development." Therefore, the UNICEF has recommended that the LDCs should go from "material growth to human potential", from "quantity of service to the sustained quality" and from "physical infrastructure to knowledge and scale". It is not the view only of the UNICEF but also of large sections of the international community.

Indeed, *Development Economics,* as Hirschman has pointed out, "is a comparatively young area of inquiry. It was born just about a generation ago as a sub-discipline of economics, with a number of other social sciences looking on both skeptically and jealously from a distance." A.K. Sen reacted to that statement by saying that Hirschman's essay that began so cheerfully turned out to be really an obituary of development economics. As a sub-discipline the same fate was predicted for economic growth as it could not meet the challenge of backwardness. In his words, "Our sub-discipline had achieved its considerable lustre and excitement through the implicit idea that I could slay the dragon of backwardness virtually by itself or, at least, that its contribution to this task was central. We now know that this is not so."

The confusion continued to persist as to which of the two concepts was a sub-discipline of the other. Whereas Hirschman maintained that economics is the sub-discipline of development, Sen argued that "the discipline of development economics does have a central law in the field of economic growth and developing countries—and that the problematique underlying the approach of traditional development economics, in some important ways, quite limited, and has not—and could not have—brought us to an adequate understanding of economic development." Unfortunately, neither is willing to give up the central role for the growth theory.

Ninth, the social costs of economic growth is now increasingly accepted as explicitly relevant to growth theory. But no growth theory has yet succeeded in integrating itself with ecology, let alone other aspects. The planners after having ravaged the economy looked for some ameliorative measures such as social forestry, protecting degraded soil and other means. But why in the first instance go for the growth measures only in terms of goods and income?

Besides, the depletion of non-renewable energy resources, there are other constraints on the continuation of crude growth model. The most important of which are: (a) environmental constraints; (b) diminishing return of economies of scale; (c) limits to the application of automation; and (d) technology qua technology. Despite the information revolution, the prospects of industrial revolution are coming to an end. Even those who believe in the possibility of resource creation recognise these limitations and therefore agree to having a realistic growth expectation.

Lester Brown produces year by year a report which fully confirms the three central theories of the Limits to Growth, namely: (a) that the short-term economic growth is being achieved at a high social long-term cost; (b) that notwithstanding all the developments and technologies, the constraints of a finite system ran supreme; and (c) that the growth based on the existing division of natural resources under the control of nation-states increases inequalities even further. Brown is not the only one to support the MIT thesis. Hundreds of ecological groups on the one hand and international reports on growing economic inequalities on the other support it.

Recently, debate on the quality of life has forced economists to agree to the simple compelling fact that growth can at best be a necessary but not a sufficient satisfying and reasonable criteria for quality of life. Beyond the satisfaction of physical needs, growth may even violate these necessary conditions also by producing negative consequences for many aspects of life.

Misleading

Power and strength of the growth theory lies in its unsocialisation and hence is misleading. In a much admired work by Srafa, production of commodities by commodities, human beings do not enter. Modern management science, industrial relations, entrepreneurial dynamics are the few sociological components which can make all the difference. Growth economists have looked upon the technology and not sociology as the main determinant. Countries which have a slow growth rate have suffered from sociological paralysis. It is the intellectuals and the ruling classes of such countries which remain preoccupied with growth which in the end leads to neurotisation. Indeed, preoccupation of growth theory supposes social pathology of the economic growth theory.

Concentration on the growth theory has played havoc with the lives of millions whose systems are organised on the basis of dual economies and with dual social and cultural structures. Development economics has focussed only on one or two types of dualism on a more comprehensive basis. Analysis of societies like India immediately brings out the multiplicity of dualistic structures on which not only the economy but the state and political power and cultural wealth are organised. We know the well known dualism between high technological enclaves and the rest of the economy in which people are underemployed. We have also seen modernisation of agriculture in a few areas and the rest suffering from subsistence agriculture. The green revolution areas have to be contrasted with vast dry farming areas which have low productivity, high labour use and low technologies. The labour may be fully employed or underemployed but is paid at less than the subsistence rate. Then, there are various kinds of rural-urban dualities or dichotomies created through the concentration of investment and social services in the urban areas. The urban poor may also be denied such services but the rural areas as a whole are given stepmotherly treatment. Uses of new technologies are creating the kind of disparities which is nothing but a new kind of dualism. The most important dualism may be cultural but it has its basis in economics.

Tenth, the syndrome of acute poverty, reasonably high rate of savings, relatively moderate level of inflation compared with other LDCs and modest growth rate has revealed serious structural crises and many vicious circles. No matter, how high the growth target planners may fix, it will be pulled down by the rest of the syndrome. A high rate of savings has been achieved with the help of gross inequalities, tax evasion, high capital-intensity of industrial output and the near financial revolution India has achieved to promote financial savings. Poverty combined with gross inequalities can produce high savings but not high income growth. The vicious circle became entrenched. The msot vicious of all circles is that poverty by becoming a barrier against inflation is also a barrier against revolution or social transformation because the poor are either too poor to revolt or are skillfully deceived by the so-called anti-poverty programme.

Growth fixation may be desirable to push the economy towards increasing savings. But once the savings rate reaches the so-called take-off stage, growth of output per unit of labour force will depend on the choice of techniques and technological progress. If the assumption of over 20 per cent rate of saving is correct, then emphasis should shift towards relevant technological choices and let the growth rate be the outcome. But if it is agreed that the growth rate can be replaced by some other variable, why technology and not employment.

Eleventh, the growth debate is also dominated either by hardcore technolgoical optimists or the prophets of doom with the balance instantaneously tilting in favour of one or the other. In both cases, there is an absence of social design. In the former, growth is expected to take care of the problem if technologies are themselves well managed. In the latter, if natural resources and environment are not protected, other things will not be able to take care of themselves. The question of human values, social obligations, relation between man and woman or between man and nature in general are not central to the debate. Most significantly, political issues involved which may require restructuring of the power system are jsut not raised.

The Contradiction

Accelerated technical progress intensifies the contradiction between the technical spheres, the biosphere and the social sphere including the international sphere. The intensification of the conflict between the techno-sphere and the bio-sphere has important economic, legal and international consequences: (a) free goods as defined by economic theory have ceased to exist; in other words, the air and water have become cost factors; (b) looked at from the aspect of constitutional law, the environmental effects of economic growth are not a national but an international problem; (c) the solution of important environmental problems require international agreements, a common budget, and international norms and verification. But there is no global economic growth theory to encompass these issues.

The recent experience of economic development tells us that there are unexpected consequences of the use of technology. We are not yet conscious of the unexpected consequences of trade expansion. The links between the two are of profound significance and leave the growth theory far behind. Trade expansion is taking place at the cost of ecology and for the solution of the latter technology is expected to come to our rescue. Even if one assumes that such a solution is possible now, there is no one to assume that this will be available in the future. The theory of the so-called "technology fix" has its serious limitations. While the ability of technology to solve certain problems has been quite impressive, there are other human problems for which there is no technology solution. On the contrary, technology itself played a major role in causing these problems. Unfortunately, these problems are still subsumed under the same growth theory.

By putting growth in the centre of economic policies, it became inevitable for every country to link growth with trade. The way the GDP was defined not only included an external sector for quantitative calculation, but it became explicitly clear that the expansion of the external sector was as important as that of the domestic economy for pushing up the growth rate. In this game, the strong economies, no matter what the trade theory said, profited at the cost of the weak economies irrespective of the fact whether trade grew rapidly or slowly. In the last few years, the competitive character of internationally economic relations has dominated the economic policies of the ADCs dragging with them the LDCs as well. The latter allowed themselves to be dragged without realising that they were bound to lose the ground to the former.

Finally, what meaning can one give to such macro-concepts as GDP, growth rate, savings rates when at least a third of income is generated in black economy and at best only partially recorded? Do we assume the same growth rate for the latter as for the former? What time frame is relevant, the very short run which does not allow income and expenditure generated in the black economy to be adjusted with the rest or the long run to allow for full adjustment? Most significantly, we seem to exaggerate the real saving rate (S/Y) because the numerator is much nearer reality than the denominator. It seems the figure of very high saving rate is exaggerated. Therefore, even in respect of mere statistical calculations the growth rate and other related variables present a distorted picture.

Enthroning Employment

Essentially, there is no dichotomy between employment and growth. It is only a question of which one takes precedence or is predetermined. In fact, if the two are kept interdependent, there will be an inducement to choose an employment rate as will increase production and thereby improve the growth rate. At the same time, with employment having the central focus, growth strategy will have to be such as to increase efficiency and productivity of the labour force.

Similarly, there should be no dichotomy bebtween employment and social justice because without employment there can be no real social justice. Those who have been advocating egalitarian and a just society have not always been alert to the need for providing full employment without which resources may be taken away from the investment pool for meeting basic needs.

No one has made the case for providing employment in countries like India in the Keynesian sense. Ironically, the situation is more neo-classical according to which the wage behaviour determines the number of unemployed. Millions remain technically employed but on very low wages. A substantial portion of employed are underemployed at less than subsistence rates. Therefore, employment by itself is no guarantee for removal of poverty just as growth is not. By and large, it is true that the poor cannot afford to remain unemployed. Besides, there are a large number of people who are self-employed. Their earnings may also be very small. The most important group which comes under the category of poverty and underemployment are small and marginal farmers. Sometimes, we ignore the fact that the figures for unemployment are not very shocking because of the way the data on poverty are collected. Therefore, a full employment strategy would require employment or self-employment at reasonable wages or earnings. In fact, poverty, underemployment, unemployment, low wages, all constitute a syndrome from which we have to get out. Therefore, the focus on employment, productivity and growth together constitute the answer and not the grwoth rate mainly.

Normally growth, productivity and employment are the three important variables which are normatively analysed. It is the objective of economic policy to improve all three in quantitative terms. All three are related, but their mutual relationships vary both in direction and value. The general experience of the last three decades in India has been that on the one hand there has been rigidity in occupational structure and distribution of labour force, on the other hand a considerable shift in the structure of income from the primary to other sectors has taken place, mostly tertiary. The growth rate has not merely concealed this divergence, but it made it difficult to provide a normative model between sectoral and occupational growth. Therefore, it is important that the growth has to be taken out in order to highlight the productivity and employment and to achieve changes in the occupational structure. It is well-known that in recent years, the elasticity in employment has declined in both industry and in manufacture and this negative factor has been concealed by the macro-growth figure.

Dethroning of the growth rate does not mean that the growth concept is not relevant. It also does not mean that it is to be traded off with some other parameters. We should not fall into the same trap as did the proponents of growth rate in making the mistake of trading off growth with other objectives and hoping that there would be an adequate trickle down towards the latter. Growth, capital formation, employment and income distribution, are all jointly determined. However, it must be recognised that by manipulation and state intervention in one parameter, the whole set of relationships will be changed. It need not necessarily be in conformity with the planner's expectations. Just as there is an intervention from the side of the planners, there may be an equal and opposite intervention from the market behaviour.

Nevertheless, since both a high and low growth rate within the possible range that India has achieved have not made a serious impact either on employment or poverty, growth has to be removed from a situation where it determines the main policies. The only other parameter which can be brought into the focus is employment. As everyone now agrees, without adequate employment, it would not be possible to change the production pattern and the standard of living.

In His Earlier Incarnation

Economists of various persuasions, right and left, have discussed poverty as a function of every variable except unemployment. The difficulty of defining unemployment in developing countries in which there is a much greater problem of under-employment makes it even more difficult to produce a policy package which would make removal of poverty a function of employment. A.K. Sen in his earlier incarnation wrote the following:

> "Poverty is a function of technology and productivity, ownership of the means of production, and social arrangements for production and distribution. To identify unemployment with poverty seems to impoverish both notions since they relate to two quite different categories of thought. Further, it can also suggest erroneous policy measures in seeking extra work for a person who is already working very hard but is poor".

However, Prof. Sen revised his view subsequently in his so-called Entitlement Theory, in which employment is one of the critical entitlements that a person can have but most of the economists still concentrate on a high growth rate as a sufficient condition for the removal of poverty. The experience of many countries did not support this hypothesis and further suggested that there is a relation of employment to income distribution. Even then it was realised that "There are very few methods of distributing income that are thought to be as effective as offering employment. It is worth inquiring why this should be so".

Integration of employment and growth objectives each taken as an independent determinant, is not possible but it is quite possible to derive one from the other. At the early stages, India's planning hovered on generation of higher economic surpluses and thus, on higher growth and let the employment growth be derived on the assumption that the latter produced low productivity levels. Despite the fact that the Mahalanobis model accepted this formualtion, the trend of growth rate until the last decade remained low, around 3.5 per cent. Only in the Seventh Plan period the growth rate has shown improvement to 5.6 per cent, though in recent years a decelertion was witnessed and according to the Reserve Bank of India, the growth in GDP in 1991-92 will only be 3 per cent. Despite claims of a high growth path in the Seventh Plan period, if one draws the balance sheet in respect of various other objectives, the results have been negative for the following reasons:

(1) Short-term growth has been achieved at the cost of long-term development prospects, particularly in terms of destruction of natural resources, future projections for population, deterioration in the quality of life, vulgarity of consumption by the middle and upper classes, rapid growth of the destitute population, and above all, the emergence of a dependency syndrome.

(2) The higher growth rate was achieved at the cost of such structural changes in GDP that reduced the share of the productive sectors such as agriculture and manufacturing and enlarged the area of the non-productive sectors. But the share of population dependent on agriculture has not changed.

(3) Disparities have accentuated in the country between social groups and classes, between rural and urban population, between wage earners and property owners, between organised and unorganised sectors and men and women.

(4) There has been a near stagnation of the per capita basic consumption goods such as pulses and sugar.

(5) Rural-urban dichotomy has become more rigid in the backward states. Neglect of socio-economic and ecological considerations have led to imbalance of incomes. There is enough data to show that their growth has been accompanied by neglect of environment and destructive exploitation of natural resources.

(6) Most significantly, whatever the rate of growth of economy, the employment growth rate has not gone up. Indeed, it has declined in the Seventh Plan. Growth of employment has declined not only in hightech capital-intensive industries, but also in agriculture and traditional crafts.

The Labour, Employment and Manpower Division of the Planning Commission sometime back brought out a disturbing trend about the decline in the growth of employment in several sectors. The most disturbing is the decline in the employment elasticity in agriculture. The table below sums up the picture for all sectors:

	Sector	Employment Elasticity	Employment Growth	GDP Growth
1.	Agriculture	0.65	2.60	4.00
2.	Mining & Quarrying	0.85	7.65	8.90
3.	Manufacturing	0.60	4.20	7.00
	— Small	0.50	5.00	10.00
	— Large	0.20	1.00	5.00
4.	Electricity	0.48	4.80	10.00
5.	Construction	1.00	8.00	8.00
6.	Transport	0.35	3.15	9.00
7.	Services	0.60	3.60	6.00
	All Sectors	0.53	3.19	6.07

However, the Division categorically stated that the growth patterns envisaged above had serious implications in investment in macro-economic and sectoral policies. It made four important recommendations: (i) The growth pattern envisaged a relatively higher rate of growth in sectors like agriculture, small and unorganised manufacturing, construction and services. Where public sector investment was likely to play an important role, the shift in pattern of public sector outlays would be required.

But to a much larger extent, policies would have to be reoriented to induce private sector investment. (ii) If the proposed thrust in agricultural growth in the lagging regions was to be pursued, modifications in the criteria for allocation of Central assistance in that sector might become necessary. (iii) A shift in the pattern of manufacturing growth in favour of small, decentralised and rural sectors would call for reorientation of the fiscal, financial and other promotional policies for preferential treatment by size, location and product lines. (iv) In order to ensure that high wage employment in certain sectors of the economy is not achieved at the cost of expansion of employment, in these and other sectors, a review of the wage trends and wage policy would be necessary.

Let me end up by invoking Mahatma Gandhi who nearly 70 years ago saw that any economic theory or technology which was inconsistent with full employment was highly immoral, economically unjustified and socially degrading. Probably, we will have to dethrone the Nehruvian approach once and for all to undo the damage of past strategies of growth fetishism.

CHAPTER 13

Towards a Demographic Disaster

The people of India are laying a seige around and against themselves. The Census Report 1991 is a disaster document. If allowance is made from both normal and abnormal under-reporting of 1991 census, the actual growth rate of population may turn out to be a little higher than that of the previous decade. Studied along with other data, the Report brutally reminds us that since China has leapfrogged, notwithstanding some recent setbacks in death and fertility rates, into a more stable balance between human and non-human resources, India will be the largest ghetto of the world in two or three decades, if population is not drastically controlled. The most depressing but not unexpected conclusion of the Report is that India has achieved nothing in this field notwithstanding the tall claims made by the Family Welfare Ministry. For quite sometimes, I have been recommending the dismantling of this Ministry and its replacement by a non-bureaucratic agency such as the Population Commission, alongwith large scale decentralisation of population control policies directed towards those who are called the so-called target groups. Both suggestions are promptly accepted and even more promptly and cynically disposed of. Probably, we will not see the truth until the disaster actually overtakes us.

China, very harshly and inhumanly, imposed a policy of one-child norm. After some successes and some failures, the norms have been relaxed. Indians cannot be easily pursuaded but our problem could be largely solved if we could "democratically impose" a two-child family norm, beginning with the non-poor to set an example for those who otherwise have nothing to lose by producing as many children as they can. China relaxed the one-child family norm not only because it proved too unrealistic and was accompanied by harsh measures, but also because they found that mixing and fusing two different policies into one, i.e. the political and the demographic into a single state policy needed a powerful cultural back-up. Culture does not obey Marxist or any other dogmatic law.

What is common between Indian and Chinese policies is that both have removed the distinction between birth control and population control. The former referes to power of women to establish control over these bodies. The latter deals with Government and other institutional policies to win control over the demographic future. Both have realised the limitations of this approach curiously. The Chinese realised it after some successes and the Indians after massive failures.

Had China been a society of free individuals, probably she too would have failed as initial conditions of both nations were similar. A partial liberalisation has already led to a partial failure. But it is quite possible that by the time China became a full-fledged democracy she might have progressed through economic development that she would have so changed the cultural environment towards fulfilling the necessary macro economic and social conditions for controlling both functions. There would be no need for them to adopt harsh measures.

Those who are criticising both her methods and temporary setbacks have been answered by Prof Scott Francis who taught in China. According to him, "Disobeying the one-child policy is tantamount to premeditated murder—the murder of a nation as well as the murder of the very children brought to birth; if we want China to develop a modern economy as well as modern civil liberties, then we must understand the necessity of the one-child policy". Can't we in India at least appreciate the fact that we are planning a national suicide?

The United Nations Population Fund Seminar which was held in May 1988 in London and in which some of world's best known experts participated came to a rather depressing conclusion which is supported by the 1991 Census. The most critical conclusion was that "India today is poised between China and Africa. The right decision could propel it rapidly towards China which is showing signs of overcoming its vast development problems. But failure to take the right step urgently could push India into Famine and despair engulfing sub-Saharan Africa".

The seminar focussed on three indicators (a) Food Production (b) Population Control and (c) Control of Environment Degradation. India at the moment, is delicately and desperately balanced between China and Africa. In other words, it is a choice between a famine some time in the future or a permanent solution to both food and population.

In 1970 India, China and Africa each produced between 160 and 200 kilograms of grains per person per year. Today, food production per person in China has been risen to 300 kilograms per person annually. In Africa, it has fallen by one fifth. In India, there has only been a marginal increase notwithstanding the success of green revolution in some States. The green revolution boosted production enough to eliminate grain imports but not enough to raise the per capita food consumption adequately.

On the other hand, China is obliged to import food notwithstanding the fact that she has twice per capita food production that of ours. This means that India can claim self sufficiency only in the market sense and not in the sense of economic access to adequate food needs.

Instructive and Disturbing

This comparison is as instructive as it is disturbing; On a comparative graph for the three regions the curve for China is much higher, but the curves for India and Africa are not markedly different. In 1987, the Indian gradient dips to ths same point as the one for Africa, at around 150 kg of grain per person per year." From this, an even more devastating conclusion was drawn. "The Seminar noted that if India faills to check population growth and recorded priorities by shifting resources into the battle to restore its soil and water base, it is more likely to follow Africa than China."

All the three regions are facing the problem of soil and environmental degradation. However, China has been able to stop further degradation. Not only that she has put nearly one-third of the land under forests and India and Africa have not, India is left with only one third of her forest land, with nearly a million hectares of forests being lost. But the deforestation, poverty, accentuation have become the two sides of the same problem. Although a large part of deforestation is the result of greedy contractors and corrupt local politicians they could not have done that to India without the support of the rulers.

China is still pursuing (though somewhat relaxed) an unflagging population policy, whereas Africa has nothing to do. India is somewhere in the middle. For instance, in Nigeria a woman bears seven children on an average. In India, it is a 4.5 and in China the current fertility rate is 2.4 births. The important point is whether or not India would be able to bring down her comparable rate which has become static since the last 15 years.

It is not a job of the demographers to tell us that the population control goes beyond the field of specialised demography. Unfortunately, the slogan of Beyond Demography has been used to sustain recriminations among those who are equal partners in the national failure. Notwithstanding the lipservice paid to the inter-disciplinary approach, there is no dialogue even among the social scientists of different disciplines, let alone a cooperative effort among them. There is also no cooperative effort among the partners engaged with the national programme. Not surprisingly, the bureaucrats easily co-opt demographers by giving them centres of population studies all of which are repeating results and co-relations discovered a million times before in India and abroad.

Why do we get surprised if the demographic trends remarkably follow the social situation? Studies on poverty have become a vast industry with vested interests in the perpetuation of poverty itself. What is not studied is that poverty is a barrier to both a social revolution and a demographic solution. If the man subjected to family planning pressures refuses to submit because of the social situation he is condemned to live in, then it is sheer arrogance to appeal for population education.

One area in which science and technology have been totally defeated in poor and densely populated countries is that of population control. A host of technical solutions are available but these cannot be pressed into service, either because of resistance from the people or indifference and arrogance of the political masters. We now have the startling discoveries of bio-science which provide probably the best solutions and yet they are not a part of any family planning programme. Those who talk about the insufficiency of research and knowledge are either *dupes* or charlatans and research has become a conspiracy between the demographers and the Government.

There can be no let-up in technological or socio-economic measures in the fields of health, education, water and sanitation. But all this require a big social commitment. Till that comes about, efforts should be concentrated on strengthening all those measures which give women larger control over their bodies instead of mechanically promoting sterlisation or improving the phoney couple protection rate.

At the 1990 meeting of the National Development Council, I reminded the Chief Ministers about their lack of concern and seriousness about population control although each paid lip service to it in his printed speeches.

Inviting Disaster

To my three suggestions for their immediate implementation there was defeaning silence. We are deliberately inviting disaster. The irony is that when the intellectuals of the developed world are trying to discover the Indian philosophy the answers to their problems, we are totally deaf to our own.

Even the normal reaction to accelerate the existing programmes has not emerged. The Government which not only spends in all kinds of wasteful activities and bureaucratic expansion has not shown concern for mobilising people for a programme which is both in the interest of the Government and the people. It is quite obvious now that without dismantling a large part of centre's functions, power and money and transfering them to lower levels of authority and inviting direct participation of the people, the demographic disaster will overtake us before even we are conscious of what has happened.

The phrase or concept Community Participation has become very popular in all fields of social development. But what it actually is or operationally signified has defied a clear definition. At one extreme, it may mean Government patronage to voluntary organisations engaged in a limited field of a harmonious area of community. At the other end, it may mean total empowerment to the political community in respect of all the resources and functions and power not only to run development programmes but also to resolve conflicts. In between these two extremes, there can be several variants of Government-community relations.

The idea of community participation began to intrude into the literature of development strategies after the Health for All, Alma-Ata declaration and the MaCnmara's World Bank strategy of Basic Needs were launched. It was quite prominently expounded in the Jakarta Conference on Family Planning in 1981 and the Mexico Conference of 1984 which read thus: "To be effective and successful, population programmes and development activities shall be responsive to local values and needs, and those directly affected shall be involved in the decision-making process at all levels. Moreover, in these activities the full participation of the community and concerned non-Governmental organisations, in particular, women's organisations, should be encouraged." The word full participation was never spelled out.

Sadly, the questions about community participation in family planning have been asked but never satisfactorily answered. There are many examples of successfull specific local NGO experiments but they are so location, family and personality specific that their larger replication has not been possible. It is difficult to determine why and how the potential beneficiaries will participate in which activity. All this depends on a whole set of social, political and economic issues which even the most well meaning bureaucrats cannot easily understand.

Community participation in any programmes means someone in higher authority giving up specified powers to some groups lower down in the heirarchy in a way that the combined power of the two is greater than the sum of these respective powers. But the problem is not really as simple as that because in any scheme of community participation or empowerment, the authorities are reluctant to shed administratively and legally acquired power, unless constitutionally required or forced by democratic pressures. The main philosophy of community participation is that such shedding of power enhances several times the power of the community. It goes without saying that in this transaction, the structure of the participatory programme, which will play a crucial role, will have to be systematically formulated within the framework of development planning.

In the case of family planning, whatever the form of community participation, it must perform the following functions or satisfy the following conditions:

First, a family planning programme to be organisationally effective must have a structure in which the Government authority or administration must be at least equal partners in decision-making with local bodies and civil institutions. The emphasis is on partnership without which the family planning programme can never succeed. Conventional bureaucratic structures taking a patronising approach to community organisation through doling out money along with setting out schedules and targets have not succeeded and will not succeed.

However, since the Government is generally hierarchically and vertically organised, the first question to be answered and which has not been answered satisfactorily is what kind of parallel organisations at each level will exist to interact with it. If one starts from the lowest rung, the question is easily answered. The village panchayats and the village functionaries will not only interact with their immediate higher bodies in order to deliver family planning services and materials to the people. They will be answerable to one another as well as to the higher authority for accountability but their mutual interaction will be the most critical level of Government-community partnership.

A village may or may not have an NGO. The local decentralised institutions act both as an explicit governmental as well as an implicit NGO in the sense that all the adults of village community i.e., Gaon Sabha, perform both functions. In fact at the local levels, the Government and the community must be so interlocked as to become one for all practical purposes. Only then will they able to undertake such functions: (1) education and awareness about the need for family planning, (2) delivery of the family planning services and (3) link up family planning programmes with other development or social programmes so as to optimise participation in all activities.

As one goes up, the interface between the administration and community will naturally become rather less direct. Even if both are well meaning, they will not be able to escape the Webrian law of bureaucractic planning and implementation become more and more bureaucratic and the degree of community participation becomes less and less even though both may come from the same class. Further on, the process of administration openly becomes more centralised. At the top of the heirarchy, centralisation is complete and interaction is with centralised political elite rather than with national level voluntary organisations which may not always be in tune with the community.

However, if the political elite and parties take up family planning seriously, some common ground between the two can be found but in most States the commitment of the political elite to family planning is rather weak. In a federal policy, the programme and power must be fully federalised. In India, the Ministry of Health and Family Welfare has acted more as a fifth wheel. Even the 1991 Census figures do not seem to have woken them up to the needs of handling over family planning to the States. The so-called national family programme remains on paper and what we have are allocation of funds and bureaucratic target setting. The community gets most distanced from the political and bureaucratic power elite. Those who are directly involved in actual family planning programme are treated as persons of lower status.

The Watch Dog

No programme of family planning can be a success unless the community mobilises local initiative and resources and acts as the watchdog of the official programme. Accountability and cost-effectiveness are two important aspects which are often ignored with the result that either services and materials do not reach the acceptors or are leaked to other uses, often illegally. The word self-reliance has ceased to have any meaning. If it is to have a meaning, it is the accountability that is crucial and this means not merely community participation but also community empowerment.

No community can mobilise local resources, motivate people, deliver the right services at the right time unless there are community leaders who are themselves motivated and are so respected that they can make Government functionaries do their job honestly and effectively. If there is widespread corruption, there can be no real interface between the government and the community leaders. Political leaders have ceased to be community leaders in the sense of ensuing delivery of goods and services meant for the people. On the other hand, field workers and community field leaders have yet to prove their credibility directly. At best, this need has been fulfilled only partially by some NGOs. The rest of the space is blank.

Another aspect is seen in terms of supervision. But who should supervise whom? So far, it is the bureaucracy which supervises and demands performance results without caring to give an account of its own performance. The scales have to be reversed if community participation is to be effective.

If community is given the right place as well as control over and access to resources and related programmes, the entire process of family planning will change from direct "birth aversion" to having "healthy babies and healthy mothers" families which will produce quick and more durable results both in the short and long run. This does not mean that birth-aversion will stop. What will happen is that community participation will ensure the success of the strategy and the siege against the people lifted.

CHAPTER 14

A Tale of Two Indias

India today is not one but two nations at all levels. Indeed, the identity of two Indias is beyond question no matter how opaque the distinction. More significantly, the relations between the two are remorselessly reaching the breaking point. It is a historical imperative that this drift must stop; otherwise like its predecessor, the two-nation theory, although not causing a new break-up of India, may unleash a multi-point civil war. Economic policies of the last four decades have enormously contributed to the making of two Indias.

But does anybody have the power to command and coerce those who would like to lay the siege against the economy? For long, the nation's leadership has remained caught in malignant wrangling. The Congress party has become an organised deception and the Opposition a disorganised self-deception. And men and women of ideas seemed to have reached the limits of intellectual possibilities.

The history of all societies is a story of change, for good or evil, through a struggle between the forces of integration and divisiveness. Only those societies and civilisations have declined or disappeared which have been unable to face the external threat or the challenge of internal divisiveness. India is pasing through a similar phase and, on balance, it appears that the forces that divide the society are getting stronger and unless this process is reversed in time, the consequences may be disastrous.

Essentially, it is the structure of the political economy and its institutions, the forces operating through them, their integrity to follow norms and rules, that finally determine whether and how the dialectical struggle between self-reliance and globalisation or between integration and disintegration mentioned above may be resolved. In retrospect, it seems that both the economic and political models and policies adopted by now the well-exposed brutally selfish Indian power elite had within itself the seeds of economic and social divisiveness. The seeds of authoritarianism also lie there. In their greed and selfishness, the elite knows how to create crises but are totally incompetent to get us out of it. Every policy they adopt puts the burden on the other India.

The analysis presented in the previous chapters is all about the siege makers, from within and without. The wealth and the power of the siege makers is intact, all the crisis notwithstanding. It is not they who are in a crisis; it is India which is in a crisis, primarily the other India that is Bharat. Had anyone noticed any serious impact on the life of the rich, the powerful and the middle classes as a result of the drastic fiscal measures introduced by the Narasimha Rao Government, taken with or without instructions from the IMF and other international financial institutions. Five years from now, there would be bigger and bigger fiscal and external deficits as there is nothing in the horizon to suggest a different structure of the economy. Five years is too long a period now. In the mean time, we are bound to see enormous convulsion, violence and corruption.

The real problem is that the Indian state is run for and by the ruling elite which itself has turned into the siege-maker, some constituents of which will keep sabotaging every reform, particularly the bureaucrats. They have mismanaged the economy or fleeced the other India which is ready to resort to violence though not yet ready for a full scale social revolution. The more serious part of the elite sees the danger of the over-blown state and can see the need for dismantling a large point of it but since the elite is itself a beneficiary expects the state to perform somehow better than before. The Left-intellectuals as ritualists had been subservient to Moscow and Beijing for decades. Others are least concerned about Swaraj, Swadeshi, self-reliance on national dignity. This is the problem for the other India. The domestic siege-makers in conjunction with the external allies will never lift the siege unless either people revolt or a more responsible, nationalist and just elite replaces them.

The Loose Talk

The loose talk about global integration of the Indian economy conceals all the sins and tragedies of the Nehruvian model. When was the Indian economy not integrated with the outside world? There was a dual economy, one part of which was integrated with the foreign companies and dependent on the World Bank and other institutions.

All those forces, classes and interests which have laid seige around the Indian economy are too powerful to be easily brushed aside. Besides, the bureaucrats and the administrative elite are so strong that no amount of external crises will put pressure on them for change. Indeed, there is a danger that the money acquired from abroad may be frittered away as before or in a different form.

If some irrefutable evidence of two Indias was needed, it could be seen in the spread of massive violence, police repression and corruption which are openly encouraged to expand with impunity and used by one India against the other. Money power, media power, and the power of misrule have made the Indian ruling elite the most pervert and malignant.

In the land of Mahatma Gandhi, such immorality and public deception should have been inconceivable. Those who ruined mother India for four decades have been christened as twiceborn or thrice-born only if they could get hold of black money. Through the double digit inflation of the last two decades, not only has an unprecedented accumulation of black money taken place, but millions of rupees have also been transferred from the poor, the unemployed and the lower middle classes to the rich. This order of transfer will now take place in about a decade. The economic system and its elite that believed falsely that they had the resilence to provide support to all kinds of political conflicts are now cracking. Indeed, if the present situation continues, the economic and political divisions may coincide to rip the society apart.

Black money not only threatens the economy through the distortions, it also distorts the pattern of investment and expenditure; it divides the society into parasites and workers and reveals an ugly contrast between the false India's hopes and real India's despair. Black money is responsible for such political divisions which as are now causing disintegration of the political system. Corruption at the top has percolated right to the bottom. What an example of decentralisation or downward mobility? A starry-eyed intellectual once suggested that in rigid societies like India, corruption should be an instrument of speeding up decision making. Now, it is obvious to everyone that corruption has become the main factor that not merely retards and delays decision making, it also divides the society into victims and victimisers in a way that the former is forced to cooperate with the latter.

Corruption is only the convergence point of institutional decay and batching of national and international perspectives that now appear as the main enemies of the people of India. Democracy has become its own enemy. Development strategy is becoming inimical by development. Foreign policy is based on decisions made outside India, creating a new dependency situation. And so on. But corruption is also knitting together the communalisation, the criminalisation, the lumpenisation, the caste-isation and unlimited pluralisation of the ethnicity of Indian politics in a way in which democracy is becoming its own enemy and development strategy inimical to development.

It was Karl Marx who systematically developed the concept of two irreconcilably conflicting social groups within a society or a nation. But his concept was based on the distinction between specific economic classes. In his version of history, all societies are political entities and are divided into two dimidate warring classes. In fact, he raised this division to the international level imagining that the globe is divided between the international proletariant and international capitalism. But the world has defied the Marxian model. No matter how sharp the internal division or revolutionary situation, the nation state has not disappeared. In fact, it was the communist states which had been more aggressive, chauvinistic and given to a new kind of imperialism, all done in the name of Marxism, which are now facing total collapse. Yet the nations of the world can be divided into two groups: those who specialise in winning and those who specialise in losing. I leave it to the judgement of the readers to decide which group India belongs to.

One cannot deny that the Indian society is also divided on class basis. But one can easily refute the claim of a two-class model. In fact, India is a multi-class society even if one were to go by the Marxian categories because India is a curious museum of practically all modes of production known in history existing side by side. If one were to list the capitalist and the proletariat in the strict Marxian sense, even the most generalised definition will keep out a vast majority of population from the calculus. Yet, the society is divided into two in which class and non-class forces combine to oppress the people.

The actual situation is that the society is broadly divided between the rich and the poor, both at once belonging to the common modes of production; but millions of poor are just pushed out of the entire economic calculus, Marxian or non-Marxian. As development proceeded, it pushed more and more people below the proverty line. The ephemeralised India has been bulging. Besides, the profile of Indian poverty is the ugliest in the world. For example, out the 360 million persons counted in this age group 15 to 59, about one-third are excluded from the category of labour force. Except for students in the age group, others, by definition, are either engaged in household work or *incapable of work*. Marginalisation of millions is pushing them below the human or sub-human level. The situation is not Marxian but Gandhian and demands a Gandhian solution.

Not only are more and more people being pushed below the poverty line every year but an equally big chunk above the so-called poverty just does more above the subsistance level. Notwithstanding official claims, in terms of basic needs and not an arbitrary poverty line, nearly forty per cent of the poor, the blind, the illiterate of the world live in India and yet we take pride in our economic and intellectual achievements. The poverty line is defined in terms of insufficient intake of calories. Nearly twenty per cent of our population has been turned into destitutes and another twenty per cent are denied all health care facilities. The stark fact is that per capita food availability (in calories) had gone down in twenty years ended 1961 from 2050 to 1949 calories. In subsequent decades, it barely regained the old level. The per capita availability of cloth, housing and health has clearly declined. Obviously, the poor keep getting less and less in terms of entitlements. This is causing a physiological and biological decay of nearly one-fourth of the nation, if not one-half. Besides, the growing social and caste inequalities have further entrenched the divide of two Indians.

Whereas half the population lives in poverty and another third a little above it, the top 15 to 20 per cent take away about two-thirds of the income, the gap between the rich and the poor is expanding further through the educational system which leaves two-thirds of India illiterate.

The Tragic Division

The most tragic division is between those who are blind or blinded and those who have wide open eyes but turn a blind eye. This is true both in fact and metaphor. One-fourth of the blind of the world live in India. Many have a chance of getting the eyesight but are not given the chance. But our brutality is reaching new depths in our deliberately blinding those whom we do not like. But, is all this less brutal than our turning a blind eye to all this and more? Do we have to be reminded that one-third of the lepers of the world also live in India?

Next to the distinction between the rich and the poor, the sharpest distinction of two Indias emerges in the field of education and health. Education has rapidly expanded and both the rich and the poor had looked upon it as a method or an instrument of improving income or social status. Some thought of it as a weapon of radical change. But the so-called mass education has proved no threat to the upper class. Indeed the system of education proved a great threat to the poor man.

A Tale of Two Indias

The Constitution provided for universal primary education in the hope that it would give a good starting point to the vast multitudes. But after four decades, we are nowhere near the target. In fact, it is the considered opinion of experts that within the prevailing poverty syndrome, we cannot achieve universal primary education because the more the poor people are pushed into the schools, still more get out of them as drop-outs.

As we move from the primary to the higher education, we find not only that education becomes more and more irrelevant, we also find that there are two systems of education: one for those who belong to the rich or to those who are highly educated and the other for the rest; in fact, it is a misnomer to call the Indian educational system as one giving true education. In fact, there are two parallel streams. It is impossible for the children coming from the under-privileged classes with a family background of no education or of poverty to make a breakthrough into that system of education which is dominated by the rich and educated. Education has brought devastating consequences to its target groups and professed victims i.e. the poor and the schedule castes.

It is this duality of the educational system that has created a problem of lack of communication between the people on account of the language barrier. We had a long tradition of many languages. But we have compounded this difficulty by imposing on it not merely an educational system which is dominated by a minority elite; we have also fallen into the trap of giving this class a language with which they can talk among' one another and not with the masses. No wonder the more the education expanded, the larger was the number of illiterates. Hardly two to three per cent of the population can communicate in the English language. The rest have either to communicate through their regional languages or not communicate at all. It has been the biggest conspiracy of the power elite to have destroyed the possibility of the emergence of a simplified language through which the masses could communicate with one another. The cultural enslavement of the educated elite to Western civilization is the single most important factor that blinds us to our problems. Intellectually eloquent compromises with silence is matched adequately only by political sychophancy.

Probably, the worst example of two Indias is found in the health sector. The position of the last two deciles of population is really precarious. Despite 170 million tonnes of food production, the calorie intake of nearly 40 per cent is inadequate and that of the bottom 20 per cent, the destitute, is such as to push them into a state of biological decay. Every time a destitute falls ill it takes a long time for him to recover. The health care system is largely for the upper classes and makes it highly expensive for the poor. The Indian medical establishment is the most perverse in the world. The health divide, with mounting demographic pressures, is the worst example of two Indias.

Never before has India been so much socially divided against itself as it is today. In fact, the situation now is worse than what existed before 1947 when the country was partitioned on communal lines. The intensity of the communal conflict is no less today than it was at that time. It is only that the numerical odds are against the Muslims, otherwise we would have to face the demand for a new partition. But more important than the communal division and conflicts are the caste and linguistic and now massive ethnic divisions which can rip up the nation at any time. Secessionist trends are slowly or fastly emerging in many places even though they are somewhat muted. But the capacity of the system to absorb them has diminished. In all riots and acts of violence it is the poor who suffer and the elite can manipulate them at will.

The Government which has callously exploited the sentiments of the minorities and weaker sections for vote catching for three decades, launched a phoney campaign for national integration. One need not doubt the honesty of individual leaders about their desire to integrate the nation. But together they are either victims of their own past or are unvenerable comic strip heroes. It is for this kind of a situation that Marx said that great individuals appear in history first as farce and second time as tragedy.

Against all historical experience, Indian economic development of the urban areas dichotomised them. Throughtout the British period, India went through a phase of de-industrialisation of its industry which had developed over hundreds of years. Over the last 40 years, India has rapidly industrialised itself in some selected sectors. But the process of de-industrialisation of several other industries has been even faster despite the so-called great preferences for the small scale.

Never before have the rural areas been denuded of the rural industries as it is today. The new technologies that have been introduced in the rural areas are limited to agriculture only and their impact has been to create many new inequalities. That is why India has got caught in a very peculiar paradox: whereas agricultural production has increased substantially and the surpluses piled up, poverty has deepened even at a faster rate. This poverty is prevalent in the agricultural sector as well as in the sector of rural industries. It is because of this that in the last 80 years, the ratio of the present labour force engaged in agriculture has stuck around 72 per cent. Old colonialism continues in new forms.

The new rural-urban dichotomy which was forced the rural areas to confine to agriculture only and urban areas to take over the rest had really created an unprecedented situation of two Indias which, at any time in the future, can generate serious clashes. This dichotomy by its very nature cannot but convulse the society. Already we have seen the emergence of farmers' agitation as an expression of this convulsion. It must be understood that all the demands of the farmers are not either justified or unjustified. What these agitations express is the structural rigidity which is sharply dividing the rural from the urban areas. India is indeed a curious case of development. Instead of integrating the various sectors of the society, it is segregating into conflicting forces.

Quite Sinister

The tie-up of business and rentiers with bureaucracy became quite sinister as both created and shared transfer payments in a highly complex entanglement with a large part of the underworld. Apart from smuggling, the scarcity of essential imports and government controlled services contributes significantly to the size and wealth of the rentier underworld which oppresses the other India.

Economic development has imposed new regional disparities over the existing ones. Economic development does create regional or intro-regional disparities. But what has happened in India is that the few States like Punjab, Haryana, Gujarat, Maharashtra and Karnataka have, over the last decade and a half, moved forward while the rest of the States have remained static or declined thus creating a major source of regional conflicts. Indeed, the rate of growth between half a dozen progressive States and the rest is so large that it is defying rational economic planning. The more resources are allocated to the backward States to bring them up to the level of the developed States, the more slowly they seem to develop.

The danger that emerges from this particular configuration of regional development pattern that the centre is becoming economically weak and the periphery is becoming strong. This situation easily generates secessionist tendencies because the periphery can be pulled away from the soft centre. Had the situation been the reverse, namely that the centre was strong and the periphery was soft, there would have been a pull towards the centre. The relatively developed States feel that the rest of India is becoming a drag on them. Opinion even in States like Maharashtra, Gujarat and Punjab which had been the bulwark of Indian nationalism is veering around the same kind of approach as was the case earlier in Tamil Nadu or Andhra Pradesh.

It is the growing alienation between the state and the civil authority that is a matter of serious concern. Corruption is the only link left between the two when the police repression is not used. The coercive power of the state is brutally used against the common people. The masses are losing faith in the system. The elements of tragedy and crime have got strangely mixed up to create an environment of hatred from which only new hatred and violence can emerge.

One-dimensional politics in a plural society or the series of monologues and Opposition's disconsoling responses are parts of the syndrome of two Indias. Obviously, the Nehrus did not believe either in the principle of the precious middle or in the

A Tale of Two Indias

law of gravity. But the terrible polarisations that are building up and the elite's iron locomotives that are bearing down on the masses will spare no one. These days, one is obliged to choose his side like a fanatic and fanatics do permanently divide societies. Nehru, Indira and Rajiv owed to themselves and to the nation to remove the dominating dark fog of mental and moral confusion. They were given the mandate this task. But they turned out be fake.

The phrase *two Indias* may sound like semantic frustration. To many, it may just appear as a scrupulous distance between classes and social groups that have appeared in all societies. To the rulers, the phrase will appear as a diaphonous myth. I do not share these insular complacencies. India's destiny will not be a foreign gift; it would have to be a hard won position for which we need a crop of leaders who can articulate the national essence from contemporary divisiveness.

In several States, most particularly those in the Hindi belt violence is not only rampant, people believe that violence has become inevitable not only to get their demands accepted but even for communicating their views to the authorities. Strikes, dharnas and petitions have lost meaning. The candidates of parties, because of their complacency and absurdities, are going to face this violent response one way or the other. Indeed, what we may witness is not a political battle but a chronicle of sinister violence. Although violence takes many forms, the most dangerous form it has taken is the criminalisation of politics. Criminals have entered almost all parties. However, there is a new coalition emerging between criminals, the police, the dominant castes and the local bureaucrats.

No matter what the outcome of local disputes, violent or not so violent, on ethnic, communal and linguistic ground, both parties to the dispute consider *the state as the enemy*. For the first time in India, the state has lost legitimacy and is the target. The distance between Delhi and the local people has widened beyond imagination and the state is identified with the local oppression and corruption of the police and the administration. The poor who though did not get much from Mrs Indira Gandhi who believed she, as a transcendental mother, wanted to do something for them. She did most to create two Indias. That is why, in their eyes, the Government is for the rich and not for the poor and is full of black cynicism.

It is natural but very disturbing that people should move away from political parties and their unsavoury remarks on manifestos and leaders are not out of place as they face repression from the police, the powerful local economic interests and the local criminals. In the heart of areas infested with communal tensions, people refused to discuss communalism because they considered it to be a secondary problem. They were more worried about the safety of their life and property and dignity of their women folk and corruption that daily added new burdens to their living.

The absence of inner party democracy and the unwillingness of party activists to go to the people to propagate the views of their parties are alienating masses of voters from them. Both the Janata Dal and Congress face this problem. But CPI(M) and BJP do not. Nevertheless, there is an upsurge of new identity—seeking through sectarian groups. The identity process has become a specific political domain (occupation, religion and ideology). This is not entirely an unhealthy development in the absence of political parties playing their true role. But, interdependence and conflict between these groups can make democracy its own antithesis unless power is decentralised.

So far, there was some hope that the developed India might regenerate itself and assume a responsible role for itself by removing the shackles and the seige put around the economy. Now there will be one section of industrial India which would be totally integrated into the international system through trade and import of capital and technology. The second India will still belong to that area of industry and trade which will remain tied to control and technological backwardness and is likely to remain stagnant so long as this sector remains within the framework of FERA.

We are Doomed

The new industrial policy which aims at globalisation of industry may further divide the society. If we accept the syndrome of two or three Indias, we are doomed. At the same time there is no social revolution around the corner or even at a reasonable distance that could break this syndrome and the siege around the economy. At best, we can prepare ourselves through that section of the elite and the masses which can press on with informed and defensive—offensive response against the imposition of the afoementioned three tier society and economy. Our response has also to be three-fold in industrial matters. This will have to be backed by a political response for demanding decentralisation of political power, debureaucratisation and even dismantling that part of the state which forms of the siege.

Although there is a genuine scepticism about the generous availability of foreign finance and technology, the Government has committed the nation to open certain areas to MNCs. The effort should be to isolate this internationally integrable area and force it to focus on exports and modernisation of capital goods. Let us call it the first sector. The second, the critical sector of large industry, is still under many regulations and depends largely on foreign collaboration but without acquiring technological independence and take-off. This sector will have to be unshackled from both internal regulations and external dependence, particularly in the consumer durables. It is here that the Government must adopt a new policy of fiscal and other discrimination. FERA must be abolished altogether and all new foreign capital must go to the first sector. Enterprises in the second sector must be allowed to compete fully and should be obliged to develop their own technology in some relation to imported technology, promote economies of scale through mergers and, like in Japan, should enjoy tax advantages that would be denied to those companies which will have any kind of foreign capital.

The third sector, loosely called the small and tiny must be fully protected. Products must be reversed for it and all other concessions except those necessary for promotion should be taken away from it. Under no circumstances, units coming in the first and second sectors be allowed to produce goods reserved for the third sector. Those already producing goods under reservation must be asked to vacate within a fixed period. Already, under broad branding many large units have diversified their activities even in unrelated fields. If these aspects are not introduced into the new economic policy of Rao-Manmohan Government, the division between two Indias will became even more rigid. The point at which this division can be undone is full employment policy.

The decline in employment growth in the economy, particularly in industry, further strengthens the augument for enlarging the labour-intensive industry, i.e. the third sector. With labour force increasing at two per cent per annum, if employment growth does not improve about that rate, the other India will suffer further. The society will see the kind of convulsions and violence that it has never witnessed before.

CHAPTER 15

New Economic Policy

I

When the new Congress (I) Government, headed by Mr. Narasimha Rao, took over it was faced with two terrifying crises of fiscal and balance of payments deficts. The crises have been building up over a decade. During the decade of '80s, there was so much fiscal anarchy and financial manipulation that it became difficult for the new government to plan on any rational basis. Three important data will be enough to give us some idea of the difficulties that Indira-Rajiv regime created for this nation. First, instead of the revenue surpluses of the first three decades, during the '80s, the surpluses turned into deficits, the total was Rs. 45,000 crores. Second, the internal and external debts became so large that the Government was burdened by a large interest-payment burden. The total internal debt is about Rs. 2,30,000 crores and the external debt mounted to come up near Rs. 1,00,000 crores. Third, the total trade deficit in this period was Rs. 65,700 crores. These figures conceal the fact that the Government was borrowing from the domestic and external markets to pay for debt incurred. The external aid was also largely used to meet the external debt obligations and trade deficits. The Government consumption expenditure had been increasing two to three times the rate of the growth of the economy. Indeed, the country faced a very deep financial crisis.

The other was the balance of payments crisis. The gap between imports and exports kept on widening over a decade so much so that export earnings were only fifty per cent of the value of imports. Besides, nearly 25 per cent of export earnings were needed for meeting external debt servicing obligations. Graver still, the foreign exchange reserves were reduced to a dangerously low level, capable of meeting only a few weeks of imports. India's external rating slipped to a very low level.

Therefore, the new Government could not go on with business as usual. Some drastic steps had to be taken. Even the worst critic of the Rao Government will have to admit that its policies has saved India from the total collapse of international confidence in the fiscal and economic system. That is not a small achievement. But the evil days has been postponed, not come to an end.

II

The package of policies announced by the Narashimha Rao Government falls into four categories: 1) Budgetary, 2) Industrial, 3) Trading, and 4) Financial. Each of these policies are related to one another one way or the other. The first lot of policies was announced in the budget for the year 1991-92 and the new industrial policy on the same i.e. 1991-92 and the new industrial policy on the same day i.e. 25th of July 1990. Trade and financial policies were announced subsequently. To fill some of the gaps left in the original statements, new policy statements are being made, creating a lot of uncertainty even though the direction is clear. An important aspect of the NEP is that post-budgetary policies reveal much greater pressures from IMF-Bank sources and are coming in phases.

First, the main features of the new Industrial policy are: (a) delicensing of all industries except 18 which come in the first schedule: (b) the removal of the MRTP Act restrictions on size and capacity creation; (c) removal of restrictions on expansion, merger, take-overs and appointment of directors; (d) raising of foreign equity investment upto 51 per cent in 34 industries; (e) automatic clearance of investment of both Indian and foreign companies under the new dispensation; (f) abolition of approval for phased manufacturing programmes to which was tied phased imprts; (g) removal of location restrictions outside the radius with one million or more of population; h) private participation to be allowed in some sectors covered so far in the core areas in public sector, such as defence, atomic energy, minerals and mineral oils; (i) removal of restrictions and compulsory registration on broad-banding facilities; j) a limited exit policy of sick units through their revaluation by the Bureau of Industrial and Financial Reconstruction (BIFR) and disinvestment upto 20 per cent of the government equity in selected units in the public sector, (k) increase in the limits on investment in the small scale units to the level of one crore of rupees each.

Second, apart from a massive downward revision of the rupee, the new trade policy comprises the following changes:

a) automatic clearance of import of capital goods in cases where foreign exchange is earned back in foreign equity or in export earnings; (b) removal of all restrictions on the import of capital goods upto two crores; (c) companies investing upto 51 per cent to be given encouragement to set up trading houses and to be treated at par with the Indian companies; (d) reduction in export subsidies (Rs. 1442 cores is againt Rs 3000 crores during 1991-92); (e) liberalisation of import of technology; (f) introduction of Exim scrips upto 30 per cent of the export earnings to be freely treated.

Third, in the field of finance, a number of measures has been taken for liberalisation that include: (a) removal of restrictions on the interest rate structure particularly on private sector debentures and bonds; (b) allowing the private sector to have mutual funds to shore up the capital market; (c) SEBI to have full control over the stock exchanges (d) the scrapping of the convertability clause by which 20 per cent of loans in excess of Rs 5 crores were converted into equities; (e) the setting up of Special Empowered Board (SEB) to negotiate with a specific type of large multi-nationals for direct foreign investment in selected areas to obtain sophisticated technologies; (f) provision of Rs. 700 crores towards strengthening the capital base of the banks and provision of Rs.

1500 crores for debt waiver scheme; (g) assurance that the Government will not intervene in the interest rate structure of the market in the name of halping the priority sectors, thus linking the cost of capital with capital worthiness; (h) tax on the interestt rates earned from deposits followed by increase in lending rates thus enhancing the whole interest rate strucuture.

Fourth, the budgetary changes include: (a) a reduction of fiscal deficit from 8.6 per cent to 6.6 per cent of the GDP; (b) reduction in subsidies particularly exports, food and fertilizers; (c) promise to reduce the non-plan expenditure which has now become almost 2 1/2 times that of the planned expenditure; (d) reduction in the depreciation rate from 33.3 per cent to 25 per cent in order to reduce capital intensity; (e) no let-up in external borrowings; (f) three schemes announced to mop up the black money.

III

Before I begin to analyse the new economic policy I would like to briefly state the background; of global economic theory and global economic environment.

The cold war which had been fully launched in 1947 had its economic side which remained submerged in political and military confrontation. In April 1947, a number of academies, mostly economists, were brought together by Fredrick Hayek who met in a remote mountain retreat of Mont Parelin, in Switzerland to meet the challenge of a trumphant Keynesian revolution which had over a decade pushed aside the classical and neo-classical liberalism. At the same time, all over the Western world, a large number of Marxists and left Keynesians joined hands to give a tough battle to the pre-war ruling intellectual paradigm in political and economic thought of the pre-Keynesian era.

Those who met at Mont Parelin set up the Mont Parelin Society-the international debating association of classical liberalism. Briefly, not only this school rejected Marxism but it also logically refuted the statist policies and focussed on socially beneficial consequences of economic freedom in open markets. The debate went on for over two decades until the economic realities and forces defeated Keynesianism by and large. The failure of the East European Communist economies and the rise of Reagan-Thatcher right-wing liberalism struck the final blow.

The foundations of Classical Liberalism and Neo-classical economics had three pillars: (1) the maximum play of market forces, (2) representative demcracy and (3) rule of law. Later on, to correct the ravages of capitalism, a fourth component, the welfare state was added. From thirties onwards, under the massive theoretical thrust of the Keynesian revolution and to meet the challenge of large scale unemployment, public investment was accepted as a strong policy option for maintaining effective demand. This was an additonal component. However, except in phases of depression and recession, in all others, the combined forces of welfare state and public investment produced perennial inflation, market distortions particularly of relative prices, and gross inefficiencies in public sector enterprises. The reaction to those developments brought back the demand for the re-introduction of full market discipline, so much so the pendulum slowly swung to the other side in the post-oil crisis. The supply side economics and Reagan-Thatcher Conservatism pushed out the public investment but not the welfare state.

As a consequence of these developments, the dominant paradigm that emerged and that is prevalent in the West and is part of World Bank-IMF thinking is composed of a four-fold political economy; (1) maximum play of market forces by privatisation, (2) representative democracy, (3) free trade and (4) welfare state. To this was added another component. The multi-polarization of global economy with dollar, yen and mark as the three major currencies forced the major economic powers to compete and compete fiercely but only within the context of a globalized economy. Thus, the dominant paradigm was extended from the nation-state to the global economy. Ideologically, all this was reduced to a free market rhetoric against the rhetoric of state planning.

In the process, the two undermined the post-war social democratic consensus on policy. The social democrats failed to understand why in the Western nations people voted for conservative parties. They still interpreted the world of the eighties and nineties in terms of the world of the thirties. They failed to comprehend the meaning of their own policies which had thrown out the "immiserisation" theories, put technology in the centre and relied far more on rapid growth of international trade than that of the economy.

The democratic principle of the rule of law has its counterpart in economics. It had two versions, one extreme and the other moderate. The extreme version took the form of a fierce and unending argument between the monetarists and the Keynesians. Reliance on monetary policy to create equilibrium rested on rules against state intervention. On the other hand, fiscal policy implied large scale state invervention for encouraging or discouraging demand through incentive, disincentives, controls and regulations and direct public expenditure. The second, the moderate view, suggested a comination of both. The reality on the ground is that despite the return of neo-classical liberalism, the fiscal-monetary combine remains thee central core of public policies. Even Reagan and Thatcher could not do away with it. Nevertheless, as Buchanan put it "a Constitutional demcoracy is constitutional, which means that rules matter". However, new state investment in industry and infrastructure has been discarded: instead, many existing state enterprises have been privatised.

A variety of forces, interests and ideas cut through this rather simplified picture of economic theorising. The most important aspect of modern societies in the "iron triangle" of three interests competing and or cooperating with one another. They are (a) political elite, (b) bureaucracy and (c) powerful interest groups. This triangular relationship of interests in contemporary democracy is underpinned by an dasymmetry of costs and benefits in the political market. Those who remain outside this triangle are denied the right to economic decision-making and are in fact obliged to remain "rationally ignorant" of the game-plan of the three interests. The state intervention through fiscal and regulatory redistribution is primarily concerned with creating equilibrium within these interest groups. But these interests have to yield to the people for the legitimisation of their ruling class character. Thus the iron-triangle prevents the full prevalence of the classical liberal economic system. That is why the resultant arrangement is more complex than what is covered in economic or political theory taken separately.

All the post-war theoretical disputations, policy thrusts and their partial reversal and structural changes could have been accommodated within the single paradigm but for two developments. First, the ideology of the Keynesian-revolution, the public investment through borrowing and the welfare state continued to increase the size of the state to an extent at which it significantly hurt the efficiency of the economy, as well as the tax-payers. In the western countries, the Government expenditure averaged about 40 per cent of the GNP. In some Nordic countries it went up to 50 per cent. This was partly achieved by a trade-off between the working class and the industrial class through the mediation of the state. This arrangement threatened to break down because of the tremendous increase in the size and income of the white collar workers and the middle classes. Besides, prolonged unemployment of nearly 10 per cent in the eighties created a new area of conflict between the employed and unemployed with the former developing a vested interest. For instance, a large sector of the former voted for Reagan, Thatcher and Kohl. To this, was added the resentment of the employed against pensioners who were financed by the state. These developments raised a demand for tax reform towards reduction in standard rates. Reagan in USA and Thatcher in U.K. set the tone and the pace, both for tax reduction and privatisation. The slogan was "get the state from the back of the people". The idea of the big state was discredited which provided new arguments for those swore by the virtues of the market mechanism.

Ironically, it was this slogan which also encouraged decentralisation and encouragement to local institutions whose power had been unnecessary usurped by the state. Although, the share of Government expenditure in GDP has not declined drastically in the West but the trend has been set in that direction. At least, one thing is sure. Never again will the state be allowed to assume functions which may involve large Government expenditure.

Just as there is a tide in the affairs of men/women and nations, similarly there is a tide in the fortuenes of ideas. We are passing through a phase which seems to be dominated by some combination of classical and neo-classical liberalism and a constrained welfare state. Marxist and Keynesian economic philosopherrs are down but not out.

On the other hand, in India a very strong assumption behind the Congress party or dynastic domination and the Nehru-Mahalonobis model was that the state would be able to determine the limits of the market, both economic and political. The market was seen as the Hobbesian concept in an atomised society in which every man's motion was opposed by every other man through competition. Market imperfections, both actual and assumed, were cited as a reason for large scale state intervention for which the economists provided a theoretical justification.

What about the political theory of the market? The Congress party's domination was the political equivalent of the economic role of the state. Although other political parties were allowed to form, the political market of voters was shrewedly manipulated by the Congress by having under its umbrella the ideological elements of all other political parties, thereby unsuring its long rule. There was no need to replace the Government or change the developments model. Precisely because the Congress Government visualised itself as a clearing mechanism of numerous competing interests, it was not realized that such an approach was bound to break down under the corruption, the rise of parasitic classes locked in combat with producing classes and inefficient allocation of resources. The Nehru-Mahalonobis Model became a victim of its own assumptions regarding politics and the role of the state. When the state failed to determine the limits of its own intervention, it could not determine the limits of pluralism or parasitism. Non-fulfilment of the objectives of state intervention provoked fierce protest and massive violence.

The New economic Policy (NEP) has to be evaluated in the context of the prevailing international economic environment, not only for India but for the whole of the third world. The first serious crisis of the third world, which had been building up for some years, erupted in 1982 when over thirty countries were unable to meet service obligations on their debts.

The first, the most important, was the defective development stretagy the benefits of which were appropriated by about top ten per cent of the population in each country. This small enclosed sector, as time went by, got more and more integrated with the advanced countries and alienated from their own people. They relied on corruption and manipulation to remain in power. There was little dent made on poverty and inequalities increased. Such a system could not have faced external shocks.

Second, of the two external shocks was the dramatic rise in oil prices, the burden of which was doubly borne by the third world. The other, more durable and less tractable, was the higher price of imports from the advanced developed countries (ADCs) which raised the prices of products and technologies as the costs of fuel imports increased.

Third, instead of reducing wasteful expenditure, the governments of these countries relied heavily on external borrowing. The IMF and theWorld Bank obliged them knowing fully well that the results could not be beneficial. But it gave this combine the opportunity to impose their conditionalities. The rate of interest was also increased sharply and remained high throughout the eighties, riaising the burden of debt and cost of repayment.

Fourth, the international prices of commodities exported by the third world collapsed, turning the terms of trade against the third world. Fifth, apart from the wrong development strategy, financial mismanagement led to a fiscal and balance of payment crisis. All these measures forced the LDCs to accept the structural adjustment demand by the IMF and the World Bank in return for new loans.

Finally, the global economy was getting integrated between the first and the second world among the ADCs and was being ruptured between the first and the second world on the one hand and the thrid world on the other. Nearly 90 per cent of the trade and 75 per cent of investment of the world take place within these countries. Their share of the global manufacture is also about 75 per cent. The rest of the world is not even treated as a proper periphery, let alone a part of the mainstream. They are treated as ghettos.

The division within the ADCs into currency or geographical groups and competition among them and even some conflicts are inevitable. But the more dominant characteristics are integration and globalisation. Whether it is the GATT, the formation of economic union, customs union or free trade area, all these acted as instruments for globalisation of the ADCs. Now the emerging trend is towards the formation of three main currency areas. Dollar, Yen and DM (or Ecu), all integrated into a global financial system backed by trillions of short and long term finance. Even if one leaves out Eastern Europe, the existing economic divisions or groups, i.e. Wester Europe, North America and East Asia have become powerful and integrated blocks, roughly comparable, each having a GNP ranging between $ 6.5 and $8 trillion. Some LDCs are being sucked into these blocks (Mexico and SE Asia region) while others are being left in the cold.

Indian leaders do not even know where India stands. The so-called talk of globalisation of the economy refers only to improving the enclave economy. There is a double rupture: rupture between India and the rest of the world and rupture between the 150 million better-off sections of the community and the rest of 700 million people of India. In both cases, India is moving towards being Darwinized both internally and externally. This is far more serious than the simple relationship of exploitation by one country by another or one group by another.

V

Dr. Manmohan Singh has promised that in two-three years' time both the balance of payment and fiscal deficits as well as the inflation rate would be brought under control. In his statement on 1st November 1991, he said when the Narsimha Rao Government took over the situation was one of total economic crisis and but for the Government using the crisis as an opportunity

New Economic Policy

to liberalise the economy and making it globally competitive, the country could have faced massive unemployment, reduction in fertilizer consumption and fall in domestic economic product. Significant economic change could be effected only through structural changes, including some in banking.

The Finance Minister has claimed many other advantages and objectives of the NEP. Whether these claims are valid or not only time will show. Therefore, it is not possible to discuss them in advance. There are others!

First, the new NEP will meet the crisis of balance of payments and fiscal deficits.

Second, it will rlease industries from the shackles of unnecessary controls and regulations which have become hurdles in the way of industrial growth.

Third, in the light of the general trend towards globalisation, the Indian economy cannot remain isolated and therefore has to be globalised to take advantage of global finances and technology.

Fourth, the Indian industries will be subjected to market dispcipline both internally and externally.

Fifth, it will encourage large scale private foreign investment through incentives provided by budgetary and non-budgetary policies.

Sixth, Indian industries instead of remaining isolated, artifically protected and divided will now be obliged to become more competitive, cost conscious and efficient. The public sector which has proved to be a drain on the national resources as it has been incurring massive losses will get rationalised either through internal efficiency or by semi-privatisation, i.e. participation of the private sector in its equities and management.

Seventh, India will be able to get long term and short term loans from the World Bank and the IMF to meet the balance of payments deficits and thus save itself from the humiliation of default.

Eighth, it shall be possible to mobilise large sums of money held by the NRIs and thus avoid dependence on the international financial system.

Ninth, a lot of black money can be mopped up through India

Investment Bonds and investments in housing without any questions being asked about the source of money.

Finally, inflation will be controlled by the introduction of fiscal reforms.

VI

The Critique

A comprehensive critique of the NEP is not possible. Moreover, as additional policies are being announced practically every day, we still do not know the final picture that would emerge when the whole package is complete. Nevertheless the major contours and directions of NEP are reasonably clear and unless some new structural changes are introduced voluntarily or under pressure of the international financial institutions, our frame of reference is limited to the policies announced so far.

There are three very important questions to which the answer must come before or at the end of the analysis. The very first question that comes to mind is: what is new in the NEP? There are three aspects of the question. Are the old paradigms and sets of policies rejected in favour of some new policies? Are these added to the latter to make a new package? Is the new policy simply the extension, accentuation and intensification of old policies? One may find elements of all three aspects in the NEP but my view is that few structural changes have been introduced and planned. NEP is more of an intensification of old policies. Certainly, there is no paradigm shift.

The Government on its part has claimed that NEP reflects both continuity and change without identifying either. We have to discover whether NEP deals or even points towards some definite direction. To me, it seems that NEP is a continuity with change in degree rather than in kind. It is the necessity of the crisis that has forced the Government to fill the gaps in the old model of continuity. A dynamic continuity is possible only with change.

The second question is about external pressures. There are two diametrically opposed opinions. The first is propagated by

New Economic Policy

the Leftists and the other by radicals that the new Government is surrendering to the international system. On the other hand, there are those who think that demands made by IMF and World Bank simply coincide with the long overdue structural change that India should have undertaken on her own. It is difficult to prove one or the other. In fact it would be shown that there is an underlying convergence of the two views.

A more important problem is about what is yet to come. Notwithstanding Government's protestations, it is becoming clear, the IMF and World Bank combine is demanding new changes in policies which it would be difficult for the Government to resist if loans are required. Earlier, it was claimed by the Government that the changes in policies were made entirely on its own initiative even if they seemed to conform to the demands of the IMF and World Bank combine. Now that fig leaf is gone. The Government has been plainly told that it must introduce new reforms if it wants external credits.

The suggested IMF—Bank reforms fall in several categories:

The first are the fiscal reforms which would imply reduction in custom duties in the new budget, introduction of new taxation measures to provide alternate sources of Government revenues, broad-based indirect taxes, rising excise duties as well as introduction of value added principles. Further reduction in subsidies particularly food, fertilizer and exports.

The second set of reforms relates to the restructuring of the public sector which will involve closing down of loss making units, putting a stop to its further expansion and some degree of privatisation of even profit earning units, in general. The Government will have to accept that what can be done by the private sector should not be done by the public sector.

The third set of demands has been made in respect of the further opening of the Indian economy to international competition. Besides reduction in customs duties, it would be required to open all sectors to foreign capital, offer concessions on protection of intellectual property rights and introduce other policies which form part of the IMF and World Bank conditionality package. Besides, the Government will have to create such conditions for an open competition which would be free from bureaucratic control. The multionals will have to be given free entry.

The fourth set of reforms would be in reducing the role of planning and changing the planning priorities. This demand has not yet come in the form and force as other demands have come but the signals have been given. Since a major part of the financing of the plan is done by banks and public financial institutions, the IFCI, IDBI and ICICI, these agencies will have to strictly conform to the international standards. There is also some talk of privatization of banks. The fifth demand is to link aid with reduction of defence expenditure.

The third question is whether liberalisation which began eleven years ago is now complete with the induction of the NEP. The answer is again in the negative because a lot of adhocism is still the hallmark of the new policy. The Government is holding back some of the components of the policy which it may reveal or forced to reveal as the impact of new policy is felt. The very fact that it has taken such a long time shows that even though the old and new policies subserve the interests of the small sector, internal contradictions and inter-elite and intra-elite conflicts of the dominant economic interests have not yet been resolved.

The right-wing critique takes us to the other extreme. Right-wing economists as well as some industrialists believe that the NEP is more cosmetic than real, particularly because corresponding changes in government implementing machinery have not been forthcoming. Although the government has dismantled some quantitative import restrictions in favour of tariffs, the custom duties on raw materials and capital goods are so high that the shift in discouraging new investment. India remains world's highest tariff imposing nation. Besides, explicit or implicit devaluation automatically increases the tariff rates. Devaluation makes imports costly and increases the cost of advalorum duty. Next it is said that the old discredited regime of import substitution which in reality became export substitution has largely been maintained. Tariff legislation which invariably was hedged around by loopholes has not under gone change. Why is the Government not liberalizing the consumer doubles sector which is claimed by the most dynamic sector? The Governments objective of achieving efficiency will fail if it remains hesistant on closing the inefficient enterprise under one kind of pressure for the other. The Government does not dare to have a matching exit policy. The Government's policy of scrapping has proved a non-starter. Subsidies are being balanced one way or the other as the consumer is being obliged to pay for them without the producers coughing up any thing. The government has not given up the dual pricing which is another hurdle in the way of the market principle. The Government expenditure and wastefulness show no sign of abatement. There is nothing in the new policy that may encourage the technical and managerial class as against the rentier or organised working class. Finally, the Government is unable to check the power of the corrupt politicians and bureaucrats. Thus the NEP is declared defective, half-hearted and without new priorities.

VI

Below are given a few important points of the critique:

First, the NEP no matter how it is evaluated, is focussed on the interests and functioning of the 150 million people and completely by-passes the interests of the rest of the 700 million people. There is nothing in the new policy which suggests that there would be a significant impact of the new policy on the latter. Even a genuine trickle down has not been suggested. The balance of payment and fiscal deficits as well as the policies suggested to eliminate them ultimately would remain the problem of 150 million people. Inflation, unemployment, social divisions and rural-urban dichotomy and similar other structural problems have received no policy direction. The crisis was created by the top 10 per cent of the population and it will be solved for them, if at all. The only countervailing measure taken is to enlarge the Public Distribution system. Indeed, but for the personal commitment of Narasimha Rao even the measures may not been taken.

However, if the new policies succeed in improving the quality of decision-making it will make the owners of the means of production efficient, remove corruption, reduce the burden of the state and its wasteful expenditure, guarantee some accountability and remove petty ideological confrontations, then, NEP may go a long way in making the Indian ruling class behave as a genuine ruling class. There have been too many inter-elite and intra-elite conflicts which have been quite unnecessary and unproductive and which also drain national resources. Controls and regulations have been used by one section of the economic elite to the disadvantage of the other, thus creating a large space for rentier activities. In a democracy, there are bound to be conflicts of interests but the function of the state is to reconcile those interests in a way as to improve the functionality of the system. The quality of the Indian ruling elite or classes have been deteriorating. Neither socialist nor capitalist values have been respected. The whole system has become corrupt and dysfunctional. Therefore, if the new NEP creates dis-incentives for corrupt practices, reduce the losses of the public sector and even reduces the size of the state and makes the Indian ruling class responsive, self-conscious and self-transcending, it would have served the purpose. Once this is achieved the stage would be set for the next plunge into changing economic and social priorities.

Second, over four decades, the Indian model of development has created what is called the dual economy. On the one hand an enclave of large urban industries based on modern technology was created both in the private and public sectors, which remain tied to foreign aid and technology. On the other hand, there was the rest of the economy of the poor which was left to fend for itself. The fiscal and trade systems were also designed to enclave the economy. The new economic policy will certainly strengthen this duality and do something wasteful also. If the new policies announced by the Narsimha Rao Government are fully implemented and if we assume that the expectations are fulfilled, then the nation would be divided not into two but three Indias. So far as there was some hope that the rich and the developed India might regenerate itself and assume a responsible role for itself by removing the shackles and the seige put around the economy. Now there would be a new sector of industrial India which would be totally integrated into the international system through trade and an import of capital and technology. The second India still belonged to that area of industry and trade which will remain tied to controls and technological backwardness and is likely to remain stagnant so long as this sector remains within the framework of FERA. Of course, there is not much for the tiny and truly small industries and even for the officially defined small scale sector.

Third, on behalf of the new industrial policy it is claimed that it will release the Indian industry from unnecessary bureaucratic shackles by reducing the number of clearances required from the Government. Mainly, there are two concessions: The proposed policy allows foreign investment up to 51 per cent to 100 per cent of equity on automatic basis subject to some restrictions on imports of capital goods. This concession is meant only for the foreigners or those who collaborate with Indian counterparts. In other words, this provision will strengthen the hold of the foreign companies on Indian industry. The second concession is for the Indian entrepreneurs and relates to technological imports. Even this proposal will now be available to both Indian and foreign companies in terms of import of capital goods up to some value of foreign exchange earned. The policy provides no defence against adverse impact on the domestic capital goods industry some parts of which have idle capacity. There is no selectivity in the policy in the sense that import should not be allowed in those cases where domestic market is in a position to supply capital goods in adequate quantity and quality. There is no appropriate industry plan with appropriate industrial mix, technological selection on the basis of priorities. All this has been done away with in the name of globalisation and rapid industrialisation. Apart from the aforementioned concessions, encouragement of domestic competition finds no place in the policy.

Even before the NEP was announced, foreign equity participation upto 74 per cent was allowed on the basis of high-tech industries. Now this has been extended to 100 per cent foreign equity in the case of export promotion with or without any technological high-tech assurance. In other words, a new liberalisation policy with regard to foreign investment is not accompanied by a well structured mechanism for compulsory transfer of technologies, particularly of designs.

Fourth, the sudden shift from import substitution to export promotion misses both the complementarity and the sequence. It also misses the need to remove massive distortions, dependency and the rutptures between the two. The basic criterion determining import-substitution and choosing industrial projects has so far aimed at saving foreign exchange in the short and medium periods. Some projections were made for long-term foreign exchange requirements but there was little consideration given to the fact that short-term gains in foreign exchange secured through setting up of all kinds of priority as well as non-priority industries either for limited export or import-substitution might lead to greater dependence on the world market and foreign capital and thus push India into a more serious external financial crisis. Of course, not all industries set up were of that kind. Quite a few, particularly the basic and raw material producing projects, had longterm beneficial effects but a still larger number did not fall in this category. Indeed, reliance on foreign collaboration and capital and technology as well as world market and world monopolies have led both to greater dependence on outside as well as greater and expanding influence of external capital on Indina industry, particularly the new industries, which were set up with the avowed purpose of creating economic independence. What came as an unintended consequence of old policy will now be accentuated as a consequence of NEP.

Fifth, a lot of protest has come from various quarters including the defenders of the status quo against the NEP for its surrender to the multinationals. The question that has to be asked is whether pre-NEP policies ensured independence in external economic policies and relations. It can be shown that all the large scale industrial units set up in the last four decades, both in the public and private sectors, remained heavily dependent on foreign aid and imported technology. Both the balance of payments and fiscal deficits jointly reflect that dependence. Budgetary support from foreign aid has been a major factor in the build-up of fiscal deficits. It has been deemed as not constituting dependence to ask for larger and larger aid from the World Bank and Aid India

Consortium. Whether the claim that the sharp depreciation of the rupee and NEP will enable India to pay for its imports by exports has yet to be tested. As such, the NEP is an accentuation of the old dependency syndrome. The drum-beaters of the Nehru-Mahalanobis model have failed to prove the independent character of that model.

India has signed 15,000 collaborations so far which cover almost every single big unit. In fact, it is difficult to count a dozen large units which have been subject to autonomous import substitution. It is both remarkable and ironic that not a single large industrial unit which is subject to licensing can be set up without some kind of foreign collaboration. To break the structural bottleneck industrial classification will have to be changed in a way as to make import substitution, export promotion, autonomous development, R&D progress, Government regulations, Government systems and subsidies as part of a new industrial paradigm or model. This is not even being thought of. If the old structure is kept intact, there is nothing in the new policy to suggest that there will be an autonomous Indian industrial structure. Besides, the industrial growth rate will remain constrained by external investment.

Sixth, ironically the emphasis on capital-intensive industries did not in the past produce technological independence. The gap between Indian and foreign technologies is the widest ever and is widening every day. Domestic R&D and designs are either killed or remain utilised precisely because of indiscriminate imports of capital goods for import substitution only. We globalised the industry all along but deliberately made it inefficient, uncompetitive, high-cost and gave it a sheltered but restrictive market. Indeed, Indian industrialisation divided the nation into two, a decision which was superimposed on the two Indias that was the creation of the colonial phase of two hundred years.

Is there anything in the NEP which will ensure the buildup of a strong autonomous technological and R&D and design base without which long term trade competitiveness can never be achieved. The failure to make headway in this direction by a soft approach on foreign collaborations and imported technology as a complete substitute for domestic technological development was the main reason for poor competitiveness of Indian goods in the world market. Indian industries and scientific establishments have remained miles apart from one another. Indian laboratories whom Nehru called temples have truly remained temples of worship. It seems to me that even within the narrower framework mentioned above and stated objectives of NEP, there is going to be no technological breakthrough. There will be larger imports.

Seventh, the Indian industry will not be able to meet the international challenge of competition unless conditions exist for good domestic competition. The Indian industry has lived so much and for so long under protection that it has become totally insensitive to the needs of quality and demand management. Almost everybody would agree on this proposition that conflicting interests and ideologies thwart efforts to create an organisational pattern for ensuring all round competition.

The main structural division of the Indian industry is between (a) handicrafts; (b) tiny; (c) small scale sector; (d) medium and (e) large industriess. By no economic logic or principle it is possible to argue that all the five must be free to compete with one another. It is strange that economists often overstretch the infant industry argument in international trade but remain oblivious of its need in domestic markets.

The first requirement of making domestic economy competitive is that each of the four sectors first became competitive within its own defined area before it is exposed to the other sector. The big business competing with handicrafts in the name of competition is a cruel joke. Since the employment coefficient is high in the first three categories, each sector must remain in internal competition until such time as the economies of scale and technological compulsions dictate otherwise. The most critical policy requirement is that the small scale sector including the tiny and handicraft should have thousands of commodities reserved for them from which the medium and large industries will be completely excluded. India is a poor country with millions unemployed. We can for a whole do with shoddy goods for domestic consumption. Except for reservation, all other concessions should be withdrawn for each sub-sector. This is the only way that both a high level of employment and competition can be guaranteed.

We have often ignored the Schumpeterian thesis that non-price or quality competition, i.e. the spread and diffusion of technology in contrast to the price factor is what 'bombarement is in comparison with forcing a door'. The Indian industry had been shut out from competition and all its problems have been pressed upon in the small scale and rural industries thus creating all kinds of contradictions, the chief being the perversity of correlation between industrial investment and industrial growth.

The new industrial policy in respect of the small scale, labour-intensive industries has completely gone in the opposite direction. Although, the SSI has been given a definition of investment upto one crore, the concessions given to it will be gradually withdrawn, with the exception of liberal credit facilities. With 24 per cent equity allowed to large domestic private sector companies and even foreign companies in the small scale sectors, the distinction between the large and small will practically disappear. This is permissible for ancilliaries, but for independent SSIs it will prove to be a disaster. No wonder, the real small scale sector has protested against the new policy. In reality what is now called small is actually medium industry of which no definition has been given. The worst sufferers will be the tiny sector and the rural industries.

Eighth, there is total confusion in both privatisation and exit policies. The latter evokes political and trade union resistance. But it should have been possible to evolve a more coherent policy with regard to privatisation. Privatissation has two meaning in economics. In one sense, it refers to the sale of government assets, particularly public sector undertakings, to individuals. In the second and more important sense, it refers to the introduction, through a joint effort between public and private sector, of conditions which insist upon public sector to follow the market principle and commercial behaviour. The NEP has accepted the first partly but has little to say anything about the latter. The triangle of interests of which the bureaucracy is the linchpin will resist the introduction of commercial principle in public sector.

They may yield on partial privatisation, particularly of loss-making units, but they are neither equipped nor find it in their interests to accept the latter. Therefore, the real problem is whether the NEP breaks or strengthens the tyranny of the *status quo*.

The implicit assumption of the NEP is that the continuing crises, in one form or another, will break the power and resistance of the bureaucracy against reforming the economy. Liberalisation started more than a decade ago and yet the process is not complete and so we do not know when, how and where will it end. There is one ugly aspect of this process. It has built into it an internal sabotage by the bureaucratic machine without the dismantling of a large part of which and its vested interests, liberalisation can be a long, agnoising, costly and frustrating exercise. There is no guarantee that the new industrial policy, like the earlier ones, will not be sabotaged. The reach of the twin-crisis of budgetary and BOP deficits to the point of bankruptcy made it possible for a partial pulling down of the resistance of the rigid bureaucracy and populist ideology of the Left. But until the package is complete, there will be uncertainty and investment decisions will be postponed. Practically, every week the Government is announcing a new set of policies. Both the investors and NRIs are following a stance of wait and see.

Ninth, the new policy is most unacceptable on the ground that it is totally silent on employment and unemployment. During the eighties, when the industrial growth increased from five per cent eight per cent, employment elasticities uniformly declined in all sectors, except the services sector. Unemployment is now becoming politically unacceptable and already leading to massive social unrest. One expected of the Government to make a clear statement on the employment objective, particularly when there is going to be a massive shift towards inviting foreign captial which will be invested only in capital-intensive industries. Modernisation and export promotion will intensify capital intensity as well as import-intensity which is also biased in favour of capital.

The new policy lumped together the need for export and wage good industry modernisation. This is the crux of the new approach.

In fact, the aim is to extend liberalisation to the entire consumer goods sector thus further ensuring de-industrialisation of rural and small scale industries which are labour intensive. Consequently, through liberalisation the entire Indian consumer goods industry will become more capital intensive and given the deficiencies of management and absence of other inputs, there is no chance of the consumer goods sector becoming internationally competitive because our advantage lies in cheap labour and not scarce capital. Instead of getting cheaper goods, the goods will be costlier and the market narrowed. Moreover, the loss of income due to the displacement of the small scale sector will be enormously high. Even if there are prospects for long term growth of capital-intensive consumer goods industries, the short and medium term losses will be enormous with incalculable social cost. Moreover, the balance of payment situation will worsen if long-term labour-intensive products are not exported.

Tenth, on the new policy there is nothing about management which can match the import of capital and technology. Where are we going to get the new entrepreneurs who will be able to break through hundreds of other controls and regulations. Modern technology, capital and efficient management and honest political environment are all necessary. The so-called technological revoluation of the developed world has so many ingredients that it will be very difficult to imagine that all those will be available to us in the short term. Since the current situation in developed countries is characterised by a shorter product life cycle along with the advantage of economies of scale buteressed by tremendous R&D efforts and venture capital that itself is internationally supported, it seems rather difficult for any late-coming third world country to find a place in this revolution.

In the various industrial policiess and their implementation, professional management was not given the pride of place though without which industrialisation and infrastructural development of the large size would not take place. Even the private sector which had to face some kind of competition, even if limited, ignored the professional management until very recently. The managing agency system militated against professional management and its abolition was accompanied by policies of sheltered market, easy foreign collaboration, import based import substitution, lack of commitment to technology development, improvement in R&D etc., all of which put a low premium on professional management. Even the multinationals which came to India followed the Indian pattern rather than Indians learning the management practices of the multinationals. Even when older countries like USA and Britain were being pushed out by the Germans and Japanese largely because of the latter's new management and technologies, we frowned on that change. We did not realise that the expanding frontiers of knowledge and technological explosion and even adaptation and innovations required a new kind of management system and practices.

A wellknown management expert and practioner, Mr. Ajit Haksar has listed five factors which have affected the management scenario: Two are of historical nature. First, the flourishing small scale sector of the pre-British period was allowed to disintegrate not only by the British period but also by the Indian government, turning producers into traders and developinga trading rather than an entrepreneurial mentality. The second was the deliberate attempt on the part of the British to retard the development of Indian industries, indeed retard the general economic growth through exploitation and making India a captive market for the British goods. The other three factors belong to the post-independence period.

'The third factor has been the predilection of the Indian psyche preferring ideas and things foreign, possibly out of the conditioning of the mind over two hundred years of colonial domination. In a sense, this has been like putting a square peg in a round hole. More than this, it has deterred the thinking to develop our own ways and approach to management despite a vast amount of relevant knowledge available from our heritage. The fourth factor has been the seller's market in most things till recently and still pertaining in many years. The fifth factor is the lingering impact of colonial rule leaving behind an undeveloped economy. Even today the vast segment of industry and commerce, at least in numbers, is entrepreneurial proprietorship in which professional management plays no part.'

Eleventh, the financial policies are designed by the Government to avoid current default in external payment and to reduce the budget deficit. These are unavoidable objectives. But the claims that the budget deficit will also result in reduction in overall fiscal deficit cannot be accepted at face value. What has been reduced by way of revenue deficit may be made up by increased new borrowing and increase in interest payment obligations. With proper statistical adjustments and removal of manipulations in the capital budget it can be shown that the overall fiscal deficit will remain at the last year's level. For instance the total interest burden will amount to Rs. 27,450 crores in 1991-92, i.e., 42 per cent of the current revenue receipts. It will go up to Rs. 35,000 crores i.e. 50 per cent of the revenue receipts in the year after.

Methodology of option exercised to reduce fiscal deficits does not inspire confidence. In the budget for 1991-92, the reduction in the budget deficit is planned to be brought about by a reduction in expenditure to the tune of Rs. 4450 crores. This is to be acheived by reducing the fertilizer subsidies by Rs. 1720 crores, reduction in loans and grants to the State Governments to the tune of Rs. 800 crores and another reduction in non-plan expenditure by Rs. 300 crores. The first item passes the whole burden to the poor by way of increased food prices. The second and third are unlikely to be achieved if one goes by past experience. There is no reason to expect that the State Governments or the Central bureaucracy will oblige the Finance Minister. There is no real shift in policies to reduce Government expenditure. There was no reason for the Government to appoint 900 new officers to the Central Services this year. They are not likely to help the economic policy in terms of cost effectiveness.

Twelveth, the balance of payments deficit is not likely to decline either, certainly not during this year or the next. Already, we are running a trade gap of Rs. 18,000 crores which is bound to go up at least by another two to three thousand crores. The favourable impact of devaluation can show results after at least two years. The prospects for global trade expansion remain rather modest in view of the projected weak growth rates of the advanced developed countries of one to two per cent in the next eighteen months. Prospects for private foreign investment too are not very bright as the global investment demand-supply gap is rather large, about $ 100 billion. The external debt which is equivalent to about 23 per cent of the GDP is likely to go up by another 3 percentage point in one year. Both the fiscal and balance of payments deficits, which was the purpose of the whole exercise of having a new economic policy are not likely to decline, for at least three years. The prices too will go their own way.

However, the objective of avoiding short-term default and arranging the second tranche of a large IMF loan may be achieved. Since the grant of a second loan requires the so-called IMF structural adjustments some components of the new policies mark a faithful response to that necessity. Whether adjustments were self-engineered or inflicted upon by the IMF need not be debated. What is to be debated is the right or wrong of the policy. An adequate defensive-offensive response is absolutely necessary.

Devaluation of the rupee has made matters worse. The confidence in the rupee has been broken. External borrowing on serveral conditionalities may help us tide over the current crisis, but a bigger crisis will most certainly overtake us because the Government is so fragile that it cannot undertake structural changes. Already there is a demand for making the rupee convertible. A few years from now, the pressures will become irresistible. The Government has hinted that after two or three years, India will voluntarily make the rupee convertible. It is not possible to discuss all the consequeneces that may flow from that step but it is worth reminding ourselves that it was during the period when the Brazalian currency was made convertible that most of the illegal transfer of capital took place. It is estimated that out of $ 70 billion external debt, about $ 40 billion has been allegedly transferred abroad.

The more the trade is liberalised and more items are brought under OGL, as was done recently, permitting imports under the Exim scrip scheme, the higher the the premium on Exim scrips. There are reports of floating fake bank certificates and frauds in

Exim scrips. The premium on Exim scrips to the level of which valuable resources will be drained away increased as imports were liberalised, thus causing a double disadvantage, rise in imports and increase in the cost of imports. Since the implicit devaluation will keep the dollar value of the rupee high, export earnings will either remain constant or increase only slowly.

Thirteenth, on the assumption that all the new policies announced go through successfully, there is a danger of stagflaton during the next two years. The industry has repeatedly voiced such misgivings. The growth of the economy has already slowed down for a variety of reasons. Stagflation will take place for the wrong reasons. They very measures, particularly new direct and indirect taxes and reduction in the depreciation rate, whcih are the right reasons to reduce fiscal deficits, will hurt industrial investment. If domestic liberalisation is speeded up, the negative impact of these measures will be partly moderated. If foreign investment rapidly increaces and exports get a boost, once again there will be some positive impact.

Whereas the unshakling and liberalising of the economy, particularly the industry, was absolutely necessary, the measures taken can achieve the stated objective over a long time. The situation, both on payments and fiscal deficits, may deteriorate before it gets better. Although confliciting claims have been made on the impact of IMF-Bank imposed structural adjustments, a recent UNCTAD monograph has pointed to the fact that in countries with large double deficits, the situation deteriorated after th socalled reforms were introduced. As far as one can see, emphasis on import-induced exports, limited external trade opportunites, devaluation of the rupee raising the cost of imports and limited supply of exportables is not likely to reduce the trade deficit. And even if the deficit is somewhat reduced, the servicing of debt will still depend upon further borrowing from abroad.

Fourteenth, the Government is making a serious mistake in raising the rate of interest. Even **Manmohan Singh** cannot deny that external pressures as much as financial crisis necessitated it. In fact the two pressures have together precluded those changes which would bring down government borrowings.

The eighties unmistably proved that when inflation remains around 10 per cent, the Bank Rate of even 12 to 13 per cent, which means the market rate of about 20 per cent will not increase savings rate. Indeed, three to four per centage points increase has lowered rather than increase the rate of savings.

About 80 per cent of India household hardly save. This opportunity cost of savings in terms of consumption is so high that no rate of interest can motivate them to save. Besides, inflation rate of 10 per cent is critical rate at which poverty ratios and subsistance levels do not change with an economic growth rate of five per cent. Indeed, prevailing levels of poverty and subsistance show further deterioration.

Although according to standard monetary theory below equilibrium interest-rate leads to capital flight, thereby reducing the availability of savings for domestic investment, it seems that continuously increasing rates during the eighties have not slowed the capital flight. There is a poor correlation between the two rates because of massive generation of block money which is the main ssource of capital flight. Trading mulpractices have no relation with changes in interests rates.

There are no data to show how the income and substitution effect work as expected. Given the relative scarcity of wealth in lagging economies, the income effect of higher rates of return should not be expected to overun the effects of substitution of more saving for less consumption now. Indeed, the consumption effect may be so large as to have no wealth effect because all the three effects are limited to those whose consumption changes positively whatever the other trade effects.

Lending rates have never been so high as now. The convertible consequences of the new interest rate policies will be (a) high cost of capital, (b) cost overruns on old projects, (c) increase in the interest burden of the debt, (d) recessionary possibilities; (e) encourage cost push inlation etc. Currently the rates of bank lending have gone up to 21 to 23 per cent, rate which will be anti-growth.

Finally, 'the NEP restss on the so-called liberalisation without definition or identification of the issues involved. Until a year back, the World Bank and IMF generally applauded India's economic achievements. But in this period the scales turned against us and the international credit rating dropped to junk pond level exposing the underlying ugly and dangerous situation. Economists have agreed that this crisis situation was the result not of the Gulf war but the consequence of the policies pursued during the eighties. Indeed, one can trace the crisis to the very structure of planning and political economy that were imposed on India by the Nehru-Mahalonobis model and its bureaucratic implementation. If India has found herself suddenly faced with a large double deficit, budgetary and balance-of-payments, which threatened to bring about the collapse of the economy, it was largely due to trends specifically set through faulty liberalisation during the eighties. During the same period, the growth rate of the economy went up by one and half percentage point. The economists went hysterical until it was found that the aforementioned double deficit sustained the growth, which now is returning to its historical trend.

Over the last few years, the World Bank had exhorted India to reinforce policies of liberalisation that started in the early eighties in order to achieve a higher growth rate, increase commercial borrowings, relax import and export restrictions for technological upgradation and competitiveness, loosen Government controls and put greater emphasis on exports. All this was recommended amidst a growing balance of payments deficits and crisis. At first, the reason given for the crisis was that India had 'fallen a victim to its very success'. Later, the blame was first put on the prevalence of controls and mismangement of the economy. How those suggestions were related to one another was never explained.

Liberalisation like regulation by itself is neither good nor bad; it depends on what it fulfils and what it negates. Many people still do not appreciate that the regime of controls and regulations of the first three decades put us neither on the path to rapid industrialisation nor on self-reliance in any significant way. On the contrary, after an initial push, it induced stagnation in the growth of the economy along with new dependency and massive corruption. Under the so-called liberal regime of the eighties, the growth rate went up by one percentage point but it produced four disastrous results of as of debt trap, (b) of more rampant corruption, (c) technological dependency, and (d) environmental degradation. There was utter confusion. Although there is a general consensus in favour of liberalisation no one has yet produced the right package of policies for it.

The real danger is that foreign trade strategies may become so demanding that they may overtake the Eighth Plan by default. The planners have a duty to restructure the domestic economy and to be deflected from the objectives of employment, gorwth and social justice. We are at the last quarter of a century away from the position where trade for the sake of trade can be a national priority as in the case of Japan and Germany. These nations first used trade for domestic enrichment and equity and then for grabbing the markets of the world.

VII

In conclusion: the twin-crises of internal and external deficits which burst on India before the Narsimha Rao Government took over was the radical outcome of the development strategy adopted and policies pursued over the lasst four decades in the name of socialism. It was indeed a cover for world'ss most inefficient and corrupt capitalism. The Nehru-Mahalonobis model of bureaucrutic planning was administered by a corrupt ruling class for its own benefits. If it pushed India into crisis after crisis it was hot surprising. In fact, it is gross under-estimation to call Indian rulers a ruling class in the proper sense of the word. It is a combination of the siege against India which had joined with the International power structure. The new economic policy is a double edged weapon which can relieve or intensify the pressures of the seige.

The decade which ended with a terrifying fiscal and balance of payment crisis, produced two contradictory responses: first, the demand for more liberalisation measures which came so far in small instalments, were partial and often negated in implementation, so much so that the business could not take full advantage. Second, the old Nehruvians and Marxists, attribute the crisis to the very principles of liberalisation, even through they are somewhat on the defensive in view of the collapse of the soviet empire. The reality is somewhere in the middle point. The culprit is not merely the pattern of liberalisation and the industrial policy but the political—business—bureaucratic nexus built that has been thwarting Indian industrialisation.

It has been the experience of the last decade that even a partial decontrol or deregulation by the Government made industry give a positve response even though the bureaucracy scuttled and made every attempt to erode the freedom given. Literally, hundreds of controls are still operating and unless several fifth-wheel departments and a large part of bureaucracy are dismantled, freedom of enterprise will still be a long way off. Nevertheless, over the years the business also learnt to get around controls and even depend on them through corruption and backdoor manipulation. The new industrial policy of the Narasimha Rao Government is bold in the sensse that it has announced several steps in one go even though it failed to tackle the power of the bureaucratess who are bound to put up tough resistance. Like the earlier liberalisation policy, the new one can also end up in an even bigger crisis if the bureaucratic blockages and sabotages are not removed.

The economic or, more specifically financial crisis is not fully autonomous. It is linked, both as a cause and effect with many other crisis. The whole society is caught with many-fold social convulsions. The NEP is a desperate plunge to meet some immediate econoic threats. It may or may not succeed. It has positive aspects which are welcome but there are many others which may deepen the crisis. If massive investment in the public sector and import substituion failed to make India self-reliant how can private sector, including foreign investers and export promotion achieve self-reliance under imposed external constraints?

The NEP which aims at globalisation of the industry may further divide the society. If we accept the syndrome of two or three Indias we are doomed. At the same time there is no social revolution around the corner or even at reasonable distance that could break this syndrome and the siege around the economy. At best, we can prepare ourselves through that section of the elite and the masses which can press on with an informed and defensive—offensive response againsst the imposition of the aforementioned three tier society and economy. Our response has also to be manifold in industrial matters. This will have to be backed by a political response for demanding decentralisation of political power, debure aucratisation and even dismantling that part of the state which forms part of the siege.

Although there is a genuine scepticism about the generous availability of foreing finance and technology, the government has committed the nation to opening certain areas to MNCs. The effort should be boisolate this internationally integrable area and force it to focus on exports and modernisation of capital goods. Let us call it the first sector. The second is the critical ssector of larger industry which is still under many regulations and which largely depends on foreign collaboration but without acquiring technological independence and take-off. The sector will have to be unshackled from both internal regulations and external dependence, particularly in the consumer durables. It is here that the Government must adopt a new policy of fiscal and other discrimination. FERA must be abolisshed altogether and all new foreign capital must go to the first sector. Enterprises in the second sector must be allowed to compete fully, should be obliged to develop their own technology in some relation to imported technology, promote economies of scale through mergers and, like in Japan, should enjoy tax advantages that would be denied to those companies which will have any kind of foreigh capital. The third sector, loosely called the small and tiny must be fully protected. Products must be reserved for it and all other concessions except those necessary for promotion should be taken away from it. Under no circumstances, units coming in the first and second sectors be allowed to produce goods reserved for the third sector. Those already producing goods under reservation must be asked to vacate within a fixed period. Already under broad banding many large units have diversified their activities even in unrelated fields.

If the new economic policies, particularly the industrial policy, are not to become the thin end of the wedge, it is important that a well-articulated defensive-offensive strategy should be prepared, debated and nationally accepted and fully implemented. How is it all to be done? At the political level, in the absence of a national government, either there should be a coalition of likeminded parties or if the minority government is to keep ruling, a national economic council of like-minded parties, economists and public men be set up in order both to create a consensus and ensure implementation. Either alternative is of paramount importance for averting the sabotage of policies by bureaucrats who dislike liberalisation and vested interests in the continuation of the prevailing system and for resisting external pressures. Several industrialists, including JRD Tata, have drawn attention to the imperative of both dismantling old and creating new implementation machinery if the new policy including that meant for the NRIs to import confidence is to be credible.

Like Japan, we need an open partnership between various elites involved in industrialisation and industry policy making. The concealed arrangements lead to corruption and misallocation of resources and licences, absence of competition etc. The Indian business should revert to the ethos of their forefathers who made the Bombay Plan. If we are to have a mixed economy capitalism, let it function as a healthy, honest system. There is no point in calling capitalism by another name, certainly not socialism. The real problem is that of corruption. Otherwise, we will continue to have the worst of both worlds and remain stuck with avoidable inter-elite conflicts that are not merely reducign surpluses, slowing down growth and generating black money and also bleeding the poor.

The NEP when implemented will lead to further centralisation whereas the need is of decentralised decision-making regarding the choice of location, technology and commodities. Therefore, more important than external liberalisation are the changes in the domestic policies towards deregulation and liberalisation in order to encourage competition. For the last thirty years liberalisation has been going on without matching deregulation. Therefore, before liberalisation is introduced or along with it the following decisions are necessary:

(a) Centralised planning should give way to decentralised planning.
(b) Bureaucratic decision-making should give way to decision-making at the enterprise level.
(c) Internationally liberalised sector should not take away too many domestic resources by starving other sectors.
(d) Non-competitive hurdles must be removed by encouraging competition within sectors rather than between sectors.
(e) Intellectual property rights must be protected along with the protection of scarce resources.
(f) If the public sector is not competitive and is also not to be allowed to invite foreign capital, then it must be allowed to be bound by the private sector to encourage domestic competition.
(g) Institutions such as the Reserve Bank and the Planning Commission shall be unchained and given full autonomy.

On the whole, the NEP has some positive and some negative aspects. Several of the reforms introduced were long overdue and therefore welcome. On the other hand, there are apprehensions that the expected results may not come about given the national and international environment. Even the industry is divided on it. The economists too are divided in their opinion though strangely enough several radical economists have supported the policy, particularly those who had been at one time employed by the World Bank or the IMF. Since the package of policy is not complete one has to wait and watch in the coming months and years. The uncertainty of the economic policy is also due to the uncertainty of the present Government. Probably the most serious problem is not to see the continuity in policy of the last four decades which brought about the crisis and dependency.

It is a sad and serious reflection on our national decision-making that decisions which we should have taken on our own are being forced on us, even though they are beneficial. The relaxation of imports of capital goods and raw materials, relaxation of certain exchange restrictions, clear notice to the public sector that it must generate its own resources and not heavily depend upon budgetary resources for expansion, reduction or removal of subsidies, pruning of government expenditure to cut down both the budgetary & fiscal deficits, the closing down of terminally sick units, World Bank's notice to the power sector to earn minimum returns or face cutting down of aid etc. are all decision we should have taken ourselves. There are a lot of other decisions about which there is still resistance from the bureaucracy and the vested interests. It is high time we seriously look into other delayed decisions before some foreign agencies takes them for us. If Rao-Manmohan team is not allowed by vested interests to take these decisions, no other team can or will in the foreseeable future.

However, whereas many new policies are continuation of the old ones in relation to basic structures and interest groups, there are three new elements in decision-making which are not entirely of our own. These were detailed in a copy of memorandum which was sent by the Government to IMF in August 1991 and remained secret for at least three months. The new constraints are entirely external and are openly imposed by IMF whereas the earlier constraints, beneficial or detrimental to our national interests, were implicitly suggested to or imposed on India by the international system and willingly or reluctantly accepted by us, ironically in the name of self-reliance.

First, the Finance Minister will have to consult the IMF before he presents the budget for the next year in respect of all the conditionalities instead of presenting his case post-facto. There may or may not be serious difference between the two with regard to growth rate, foreign exchange rates, inflation, fiscal deficits, etc. but surely on methods of achieving macro balances and achieving related objectives, there could be serious differences. Therefore, secondly, the Government agreeing to IMF's medium-term structural adjustment conditions can damage India, even if there is no disagreement on long term objectives. Third, and this is where real sovereignty has been compromised, economic programme will be monitored quarterly by the IMF. Several reviews, at least three, will be conducted in a year and the loans agreed upon will be released in instalments after each review, depending upon quarterly performance.

Revenue Capital Receipts and Expenditure of the Centre—1980-81 to 1991-92

(Rs crores)

Items	1980-81	1981-82	1982-83	1983-84	1984-85	1985-86	1986-87	1987-88	1988-89	1989-90	1990-91 (R.E.)	1991-92 (B.E.)
1.	2.	3.	4.	5.	6.	7.	8.	9.	10.	11.	12.	13.
1. Revenue Receipts	11937	14631	17036	19711	23466	28035	33083	37037	43591	52296	57381	67529
2. Capital Receipts	7642	8360	10995	14406	16421	19314	21572	25408	29878	30018	38564	38174
3. *Total Receipts*	*19579*	*22991*	*28031*	*34117*	*39887*	*47349*	*54655*	*62445*	*73469*	*82314*	*95945*	*105703*
4. Revenue Expenditure	14410	15408	18742	22251	27691	33924	40860	46174	54106	64208	74966	81383
(a) Plan	2379	2999	3789	4594	5679	6907	8216	9907	11119	12071	14020	17068
(b) Non-Plan	(12031)	12409	14953	17657	22012	27017	32644	36267	42987	52137	60946	64315
5. Capital Expenditure	7646	8975	10945	13283	15941	18741	22056	22087	25005	28698	31751	32039
(a) Plan	N.A.	7251	8124	9444	10927	12947	14780	14302	15033	16330	15936	16657
(b) Non-Plan	N.a.	1724	2821	3839	5014	5794	7276	7785	9972	12368	15815	15382
6. *Total Expenditure*	*22056*	*24383*	*29687*	*35534*	*43632*	*52665*	*62916*	*68261*	*79111*	*92906*	*106717*	*113422*
(a) Plan	8994	10250	11913	14038	16606	19854	22996	24209	26152	28401	29956	33725
(b) Non-Plan	13062	14133	17774	21496	27026	32811	39920	44052	52959	64505	76761	79697
7. Revenue Deficit	2473	777	1706	2540	4225	5889	7777	9137	10515	11912	17585	13854
8. Budgetary Deficit	2477	1392	1656	1417	3745	5316	8261	5816	5642	10592	10772	7719
9. Non-Plan Rev. Exp as % of Total Expenditure	54.55	50.89	50.37	49.69	50.45	51.30	51.89	53.13	54.34	56.12	57.11	56.70
10. Plan Exp. as % of Total Expenditure	40.78	42.04	40.13	39.51	38.06	37.70	36.55	35.47	33.06	30.57	28.07	29.73

Note: Figures in parentheses is R.E. as actuals are not readily available.
Source: Budgets of the Central Government

Selected Indicators

	1985-86	1986-87	1987-88	1988-89	1989-90 (G.E.)
1.	*2.*	*3.*	*4.*	*5.*	*6.*

— Rs crores at current price —

	1985-86	1986-87	1987-88	1988-89	1989-90 (G.E.)
Gross Domestic Product (GDP) at market prices	261920	291974	332616	394992	442769
Net National Product (NNP) at market prices	234254	260346	296311	353283	394620
Household Savings —Gross (GHHS)	37999	40696	54718	67439	78913
Household Savings —Net (NHHS)	27081	28590	41182	52109	61605
Household Savings —Net Financial (NFHHS)	18514	22992	26022	28646	39386
Public Sector Savings —Net (NPSS)	-2931	-5096	-7680	-9782	-13182
Net Dom. Private Corporate Sector Savings (NDPCS)	1277	421	373	1881	1947
Gross Domestic Capital Formation (GDCF)	57898	60093	74386	94432	106501
Gross Domesic Savings (GDS)	51664	53738	67561	83298	95917
Net Inflow (GCF-GCS)	6234	6355	6825	11134	10584
Aggregate Net Investment/ Capital Formation (NCF)	31661	30270	40700	55342	60971

— Per Cent —

	1985-86	1986-87	1987-88	1988-89	1989-90 (G.E.)
GHHS/GDP	14.51	13.94	16.45	17.07	17.82
NHHS/NNP	11.56	10.98	13.90	14.75	15.61
NPSS/NNP	-1.25	-1.96	-2.59	-2.77	-3.34
NFHHS/NNP	7.90	8.83	8.78	8.11	9.98
NDPCS/NNP	0.55	0.16	0.13	0.53	0.49
Net Inflow/NNP	2.66	2.44	2.30	3.15	2.68
NCF/NNP	13.52	11.63	13.74	15.67	15.45

Q.E. = Quick Estimates
Source: C.S.O. *National Accounts Statistics*

Budget Deficit of the Centre and States: 1980-81 to 1990-91

Years	Rs Crores			As Percent of GDP	
	Centre	States	Total	Centre	Total
1.	2.	3.	4.	5.	6.
Sixth Plan					
1980-81	2576	874	3450	1.89	2.54
1981-82	1392	1127	2519	0.87	1.58
1982-83	1655	694	2349	0.93	1.32
1983-84	1417*	718	2135	0.68	1.03
1984-85	3745	1360	5105	1.62	2.21
Total	10785	4773	15558	1.18	1.70
Seventh Plan					
1985-86	4937	1498	3439	1.88	1.31
1986-87	8261	889	9150	2.83	3.13
1987-88	5816	-312	5504	1.75	1.66
1988-89	5642	-540	5102	1.43	1.29
1989-90	10592	-1575	33673	2.04	1.95
Total	35248	-1575	33673	2.04	1.95
1990-91 (R.E.)	10772	-514	10258	2.18	2.08

*Excluding taken over deficit of the States of the order of Rs 399 crores.

Table I

Macro Economic Aggregates and
(At 1980-81

Year	GDP at Factor Cost	CFC	NDP at Factor Cost	Indirect Taxes Less Subsidies	GDP at Market Prices	NDP at Market Prices	GNP at Factor Cost	NNP at Factor Cost	GNP at Market Prices	NNP at Market Prices
1	2	3	4	5	6	7	8	9	10	11
1950-51	42871	2190	40681	2868	45739	43549	42644	40454	45512	43322
1951-52	43872	2287	41585	3295	47167	44880	43730	41443	47025	44738
1952-53	45117	2389	42728	3260	48377	45988	45002	42613	48262	45873
1953-54	47863	2475	45388	3498	51361	48886	47768	45293	51266	48791
1954-55	49895	2639	47256	4020	53915	51276	49750	47111	53770	51131
1955-56	51173	2831	48342	4590	55763	52932	51119	48288	55709	42878
1956-57	54086	3047	51039	4770	58856	55809	54002	50955	58772	55725
1957-58	53432	3253	50179	5314	58746	55493	53304	50051	58618	55365
1958-59	57487	3442	54045	5568	63055	59613	57311	53869	62879	59437
1959-60	58745	3659	55086	6042	64787	61128	58434	54775	64476	60817
1960-61	62904	3930	58974	5159	68063	64133	62532	58602	67691	63761
1961-62	64856	4191	60665	5860	70716	66525	64359	60168	70219	66028
1962-63	66228	4492	61736	6723	72951	68459	65657	61165	72380	67888
1963-64	69581	4790	64791	7932	77513	72723	69006	64216	76938	72148
1964-65	74858	5179	69679	8404	83262	78083	74121	68942	82525	77346
1965-66	72122	5604	66518	9179	81301	75697	71338	65734	80517	74913
1966-67	72856	6003	66853	8163	81019	75016	72092	66089	80255	74252
1967-68	78785	6326	72459	8505	87290	80964	77845	71519	86350	80024
1968-69	80841	6660	74181	9594	90435	83775	79945	73285	89539	82879
1969-70	86109	6972	79137	10246	96355	89383	85149	78177	95395	88423
1970-71	90426	7254	83172	10936	101362	94108	89465	82211	100401	93147

Population (1980-81 Series)
Prices)

(Rs crores)

PFCE in Domestic Market	GFCE	GDCF	NDCF	Per Capita GNP at Factor Cost	Per Capita NNP at Factor Cost	Rate of GDCF (Per Cent)	Rate of NDCF (Per Cent)	Annual Growth in GNP at Factor Cost	Annual Growth in NNP at Factor Cost	Annual Growth in Per Capita NNP	Population in Millions
12	13	14	15	16	17	18	19	20	21	22	23
36127	2522	6705	4515	1187.9	1126.9	14.7	10.4	—	—	—	359
37778	2548	7852	5565	1198.1	1135.4	16.6	12.4	2.5	2.4	0.8	365
38915	2551	5174	2785	1209.7	1145.5	10.7	6.1	2.9	2.8	0.9	372
41307	2583	5906	3431	1260.4	1195.1	11.5	7.0	6.1	6.3	4.3	379
42455	2598	6549	3910	1288.9	1220.5	12.1	7.6	4.1	4.0	2.1	386
42897	2670	9142	6311	1300.7	1228.7	16.4	11.9	2.8	2.5	0.7	393
44860	2856	11755	8708	1346.7	1270.7	20.0	15.6	5.6	5.5	3.4	401
44007	3216	11459	8206	1303.3	1223.7	19.5	14.8	-1.3	-1.8	-3.7	409
47863	3330	9441	5999	1371.1	1288.7	15.0	10.1	7.5	7.6	5.3	418
48265	3390	10152	6493	1371.7	1285.8	15.7	10.6	2.0	1.7	-0.2	426
51765	3573	12348	8418	1440.8	1350.3	18.1	13.1	7.0	7.0	5.0	434
53209	3836	11325	7134	1449.5	1355.1	16.0	10.7	2.9	2.7	0.4	444
53765	4629	13189	8697	1446.2	1347.2	18.1	12.7	2.0	1.7	-0.6	454
55838	5733	13876	9086	1487.2	1384.0	17.9	12.5	5.1	5.0	2.7	464
59274	5939	15189	10010	1563.7	1454.5	18.2	12.8	7.4	7.4	5.1	474
58741	6516	16888	11284	1470.9	1355.3	20.8	14.9	-3.8	-4.7	-6.8	485
59831	6572	17751	11748	1456.4	1335.1	21.9	15.7	1.1	0.5	-1.5	495
63038	6705	16227	9901	1538.4	1413.4	18.6	12.2	8.0	8.2	5.9	506
64885	7073	15296	8636	1543.3	1414.8	16.9	10.3	2.7	2.5	0.1	518
67851	7764	17581	10609	1609.6	1477.8	18.2	11.9	6.5	6.7	4.5	529
71522	8492	18928	11674	1653.7	1519.6	18.7	12.4	5.1	5.2	2.8	541

Contd....

1	2	3	4	5	6	7	8	9	10	11
1971-72	91339	7606	83733	11864	103203	95597	90281	82675	102145	94539
1972-73	91048	8006	83042	11518	102566	94560	89997	81991	101515	93509
1973-74	95192	8385	86807	10404	105596	97211	94395	86010	104799	96414
1974-75	96297	8769	87528	10558	106855	98086	95885	87116	106443	97674
1975-76	104968	9227	95741	11707	116675	107448	104660	95433	116367	107140
1976-77	106280	9743	96537	12468	118748	109005	105996	96253	118464	108721
1977-78	114219	10233	103986	13074	127293	117060	113903	103670	126977	116744
1978-79	120504	10836	109668	14134	134638	123802	120302	109466	134436	123600
1979-80	114236	11442	102794	13336	127572	116130	114379	102937	127715	116273
1980-81	122226	12087	110139	13586	135812	123725	122571	110484	136157	124070
1981-82	129600	12788	116812	15011	144611	131823	129639	116851	144650	131862
1982-83	133469	13595	119874	16464	149933	136338	132853	119258	149317	135722
1983-84	144310	14469	129841	16682	160992	146523	143306	128837	159988	145519
1984-85	149966	15448	134518	17056	167022	151574	148789	133341	165845	150397
1985-86	157348	16342	141006	20082	177430	161088	156147	138805	176229	159887
1986-87	163924	17360	146564	21983	185907	168547	162188	144828	184171	166811
1987-88	170716	18411	152305	23637	194353	175942	168378	148967	192015	173604
1988-89	187725	19343	168382	24915	212640	193297	185543	166200	210458	191115
1989-90	197419	20439	176980	25891	223310	202871	195237	174798	221128	200689

Source: Central Statistical Organisation, *National Accounts Statistics,* March

CFC   Consumption of Fixed Capital
PE   Provisional Extimate
GDCF   Gross Domestic Capital Formation
GFCE   Government Final Consumption Expenditure

NDCF   Net Domestic Capital Formation
NDP   Net Domestic Product
NDS   Net Domestic Saving
NNP   Net National Product
PFCE   Private Final Consumption Expenditure
QE   Quick Estimate.

12	13	14	15	16	17	18	19	20	21	22	23
73206	9369	19899	12293	1629.6	1492.3	19.3	12.9	0.9	0.6	-1.8	554
73647	9402	18623	10617	1587.2	1446.0	18.2	11.2	-0.3	-0.8	-3.1	567
75654	9305	23664	15279	1627.5	1482.9	22.4	15.7	4.9	4.9	2.6	580
75747	8875	21333	12564	1616.9	1469.1	20.0	12.8	1.6	1.3	-0.9	593
80063	9799	22014	12787	1724.2	1572.2	18.9	11.9	9.2	9.5	7.0	607
82165	10576	24175	14432	1709.6	1552.5	20.4	13.2	1.3	0.9	-1.3	620
88706	10898	26311	16078	1796.5	1635.2	20.7	13.7	7.5	7.7	5.3	634
94041	11706	31768	20932	1856.5	1689.3	23.6	16.9	5.6	5.6	3.3	648
91379	12424	28401	16959	1722.6	1550.3	22.3	14.6	-4.9	-6.0	-8.2	664
99083	13084	30867	18780	1805.2	1627.2	22.7	15.2	7.2	7.3	5.0	679
103574	13663	32237	19449	1868.0	1683.7	22.3	14.8	5.8	5.8	3.5	694
106604	15075	30530	16935	1873.8	1682.1	20.4	12.4	2.5	2.1	-0.1	709
114606	15750	32115	17646	1979.4	1779.5	19.9	12.0	7.9	8.0	5.8	724
119071	16983	31646	16198	2013.4	1804.3	18.9	10.7	3.8	3.5	1.4	739
123435	18924	37574	21232	2068.2	1851.7	21.2	13.2	4.9	4.8	2.6	755
129151	20848	36565	19205	2106.3	1880.9	19.7	11.4	3.9	3.6	1.6	770
134972	22783	38884	20473	2144.9	1910.4	20.0	11.6	3.8	3.5	1.6	785
146306	23928	47121	27778	2319.3	2077.5	22.2	14.4	10.6	11.2	9.2	800
151974	25044	48605	28166	2392.6	2142.1	21.8	13.9	5.2	5.2	3.1	816

1989, June 1989, March 1990 and Press Notes dated February 19, 1991.

GFCF   Gross Fixed Capital Formation
GDP   Gross Domestic Product
GDS   Gross Domestic Saving
GNP   Gross National Product

Table II

Growth of Real Domestic Product in the Indian Economy: 1950-51 to 1989-90
(At Constant 1980-81 Prices)

(Amounts in Rs. crores)

	Year	Gross Domestic Products (GDP) at		Net Domestic Product (NDP) at	
		Factor Cost *(1a)*	Market Prices *(1b)*	Factor Cost *(2a)*	Market Prices *(2b)*
By Plan Periods					
Pre-Plan Year	1950-51	42,871	45,739	40,681	43,549
Terminal Year of the First Plan	1955-56	51,173	55,763	43,342	52,932
Terminal Year of the Second Plan	1960-61	62,904	68,063	58,974	64,133
Terminal Year of the Third Plan	1965-66	72,122	81,301	66,518	75,697
Base Year for the Fourth Plan	1968-69	80,841	90,435	74,181	83,775
Terminal Year of the Fourth Plan	1973-74	95,192	105,596	86,807	97,211
Terminal Year of the Fifth Plan	1978-79	120,504	134,638	109,668	123,802
Base Year for the Sixth Plan	1979-80	114,236	127,572	102,794	116,130
Terminal Year of the Sixth Plan	1984-85	150,469	167,525	135,021	152,077
Terminal Year of the Seventh Plan	1989-90	197,419	223,310	176,980	202,871
By Decades					
	1950-51	42,871	45,739	40,681	43,549
	1960-61	62,904	68,063	58,974	64,133
	1970-71	90,426	101,362	83,172	94,108
	1980-81	122,427	136,013	110,340	123,926
	1989-90	197,419	223,310	176,980	202871

Source: National Accounts Statistics, (1991) by the Central Statistical Organisation, Department of Statistics, Ministry of Planning, Government of India.

Table III

1. National Income and Productions

1.2 Gross National Product and net National Product (i.e. National Income) (Annual Growth Rates)

Year	Gross National product at factor cost		Net National product at factor cost		Per Capita Net National Product	
	At Current Prices	At 1980-81 Prices	At Current Prices	At 1980-81 Prices	At Current Prices	At 1980-81 Prices
1	2	3	4	5	6	7
Annual Compound Growth Rates						
First Plan (1951-56)	1.7	3.7	1.3	3.6	-0.5	1.7
Second Plan (1956-61)	9.4	4.1	9.2	3.9	7.1	1.9
Third Plan (1961-66)	9.5	2.7	9.3	2.3	6.9	0.1
Three Annual Plans (1966-69)	11.6	2.4	11.7	2.2	9.3	-0.1
Fourth Plan (1969-74)	10.9	3.4	10.9	3.3	8.4	0.9
Fifth Plan (1974-79)	10.6	5.0	10.3	4.9	7.9	2.6
Annual Plan (1979-80)	9.5	-4.9	8.3	-6.0	5.7	8.2
Sixth Plan (1980-85)	15.1	5.5	14.9	5.4	12.5	3.2
Seventh Plan (1985-90)	13.6	5.5	13.4	5.5	11.2	3.4

Note: Based on data in Table 1.1

Table IV

Estimates of Domestic Saving and Investment
(At Current Market Prices)

(In percentages)

Sl. No.	Sector/Item	Fiscal Years		
		1984-85	1985-86 *(Provisional)*	1986-87 *(Quick Estimates)*
1	2	3	4	5
1.	Net household sector's saving to NNP of which:	15.6	15.6	17.0*
	Saving in financial assets	9.3	8.7	10.1
2.	Net public sector's saving to NNP	0.8	0.7	0.7*
3.	Net domestic private corporate sector's saving to NNP	0.5	0.5	0.5
4.	Total net domestic saving to NNP (1+2+3)	16.9	16.8	18.2
5.	Inflow of foreign resources to NNP	1.5	2.5	2.0
6.	Aggregate net investment to NNP (4+5)	18.4	19.3	20.2

* In the absence of the CSO's estimate of investment in physical assets by the households and the estimate of saving of the public sector, the rate of the previous year is assumed for this year also, which would undergo exchange when CSO releases its Quick Estimates in January 1988.

Table V

Incremental Gross Capital Output Ratios

S.No.	Period	Capital Output Ratios					
		New Series			Old Series		
		Using Avg. Gr. of GDP	Using Comp. Gr. of GDP	Using Semi Log	Using Avg. Gr. of GDP	Using Comp. Gr. of GDP	Using Semi Log
	(1)	(2)	(3)	(4)	(5)	(6)	(7)
1.	1951-52 to 1955-56	2.95	2.96	2.56	3.03	3.04	2.65
2.	1956-57 to 1960-61	3.40	3.44	3.58	3.66	3.71	3.78
3.	1961-62 to 1965-66	5.43	5.57	4.53	6.20	6.42	5.26
4.	1966-67 to 1970-71	3.43	3.45	2.99	3.55	3.58	3.11
5.	1971-72 to 1975-76	5.80	5.90	5.26	5.88	6.01	5.55
6.	1976-77 to 1980-81	6.70	6.95	7.56	6.57	6.81	6.96
7.	1981-82 to 1985-86	4.12	4.14	4.14	4.72	4.73	4.67
8.	1985-86 to 1989-90	4.04	4.06	3.76			

Source: Computed from National Accounts Statistics (New Series) 1950-51 to 1979-80, 1989 and National Accounts Statistics 1990. National Accounts Statistics January, 1987.

Source: S.R. Hashim, Forty Years of Indian Economy: Structure and Dimensions of Growth, Vera Anstey Memorial Lecture, 73rd Annual Conference of the Indian Economic Association (December 28-30, 1990).

Table VI

Gap between Gross Domestic Savings and Gross Domestic Capital Formation

	GDS	GDCF	Difference(%)
1970-74	17.6	18.5	0.9
1975-79	22.5	22.0	-0.5
1980-84	20.3	21.7	1.4
1985-87	20.3	24.0	3.7

Table VII

Trends in Sector-wise Net Domestic Product (NDP) at Factor Cost at 1980-81 Prices

Year	Total Primary Sector	Total Secondary Sector	Total Tertiary Sector	Total Net Domestic Product
1	2	3	4	5
1980-81	45,395 (41.2)	25,300 (23.0)	39,444 (35.8)	1,10,139
1981-82	48,149 (41.2)	27,232 (23.3)	41,431 (35.5)	1,16,812
1982-83	47,369 (39.5)	28,313 (23.6)	44,192 (36.8)	1,19,874
1983-84	52,389 (40.3)	31,113 (24.0)	46,339 (35.7)	1,29,841
1984-85	52,249 (38.8)	32,984 (24.5)	49,285 (36.6)	1,34,518
1985-86	52,326 (37.1)	35,430 (25.1)	53,250 (37.8)	1,41,006
1986-87	51,595 (35.2)	37,994 (25.9)	56,975 (38.9)	1,46,564
1987-88	51,815 (34.0)	40,029 (26.2)	60,461 (39.7)	1,52,305
1988-89	61,073 (36.1)	43,141 (25.5)	64,717 (38.3)	1,68,931
Average percentage of share (1980-81 to 1988-89)	38.2	24.6	37.2	

Note: Figures in brackets are percentages to Net Domestic Product (NDP) at factor cost.

Table VIII

Distribution of Workers by Sectors

	Industry Division	1972-73	1977-78	1983	1987-88
	Primary Sector	74.44	72.8	77.0	66.2
1.	Agriculture	74.0	72.3	68.4	65.5
2.	Mining & Quarrying	0.4	0.5	8.6	0.7
	Secondary Sector	10.8	11.7	13.1	14.8
3.	Manufacturing	8.8	9.8	10.6	10.8
4.	Electricity, Gas & Water Supply	0.2	0.2	0.3	0.3
5.	Construction	1.8	1.7	2.2	3.7
	Tertiary Sector	14.6	15.4	17.4	18.4
6.	Trade	5.0	5.8	6.2	6.9
7.	Transport	1.8	1.9	2.4	2.5
8.	Financing, Real Estate Insurance & Business services	0.5	0.5	0.7	0.8
9.	Community, Social & Personal Services	7.3	7.2	8.1	8.2
	Total	100.0	100.0	100.0	100.0

Note: Where the total do not add upto 100.0 it is due to some minimal inadmissible categories.

Source: "Employment Past Trends and Prospects for 1990s" Working Paper, Planning Commission, New Delhi, May 1990, page 3, Table 6.

Table IX

Percentage Distribution of Population in different Expenditure Groups of Households in 1988-89 (NSS 44th Round)

Monthly per capita expenditure classes (Rs.)	*Rural*	*Urban*
1	*2*	*3*
00-110	20.26	6.18
110-215	50.40	34.70
215-385	22.67	34.33
385 and above	6.66	24.79
Not recorded	0.01	0.00
All Classes	100.00	100.00

Source: 'Sarvekshana' Vol. XIV No. 3 Issue No. 46 Jan. - March 1991.

Table X
Growth Rates in Value Added at 1980-81 Prices by Industry Groups

Industry group	Weight in gross value added in registered manufacturing in 1988-89 at 1980-81 prices	GROSS VALUE ADDED (Rs. crores)										Annual growth rate % (Semi-log) basis
		1980-81	1981-82	1982-83	1983-84	1984-85	1985-86	1986-87	1987-88	1988-89		
1.	2.	3.	4.	5.	6.	7.	8.	9.	10.	11.	12.
I. Registered Manufacturing												
food products	8.85	845	1116	1506	1679	1694	1776	1816	1989	2120		10.23
beverages, tobacco, etc.	1.97	246	261	270	497	420	367	410	435	472		8.10
cotton textiles	6.43	1567	1316	1235	1535	1444	1661	1828	1591	1540		2.30
wool, silk, etc.	3.35	443	535	563	662	698	810	773	744	803		7.26
jute textiles	1.55	324	293	273	203	198	215	359	322	372		2.43
textile products	1.56	132	165	185	194	231	174	187	255	373		9.37
wood, furniture, etc.	0.49	71	71	72	91	87	90	100	111	118		6.93
paper & printing etc.	3.55	526	575	530	610	716	692	832	825	850		6.95
leather & fur products etc.	0.65	77	94	104	126	147	119	116	162	155		7.95
rubber, petroleum etc.	8.23	614	577	880	930	1074	1099	1573	1823	1971		17.05

Indian Economy Under Siege

1.	2.	3.	4.	5.	6.	7.	8.	9.	10.	11.	12.
chemicals etc.	16.58	1854	2203	2281	2808	2919	3045	3105	3417	3971	8.80
non-metallic products	5.16	474	508	636	722	913	970	922	1058	1236	12.51
basic metal industries	8.06	1556	1735	1601	1736	1733	1906	1722	1806	1931	2.06
metal products	2.60	363	368	372	417	445	478	502	604	622	7.57
non-electrical machinery tools and parts	7.76	885	1053	1127	1269	1536	1563	1445	1613	1858	8.60
electrical machinery	8.39	918	943	1212	1276	1607	1359	1480	1921	2009	10.02
transport machinery	7.26	1007	1140	1283	1401	1547	1377	1641	1539	1739	6.11
other manufacturing	6.12	456	548	677	790	945	1165	1250	1360	1465	16.20
repairing services	1.42	182	182	219	251	273	259	284	313	341	8.13
Total-I	100.00	12540	13683	15026	17197	18627	19125	20345	21888	23946	8.16
II. Mining & quarrying		1887	2141	2387	2451	2486	2623	2978	3078	3339	6.68
III. Electricity, Gas & Water supply		2070	2264	2415	2588	2863	3099	3422	3690	4127	8.88
IV. Industry (NDP)		21084	22921	24469	26723	28316	29425	31730	33898	36921	6.95

Source: CSO: National Accounts Statistics

Table XI

Growth Rates in Value Added at 1980-81 Prices by Industry Groups

Industry group	Weight in gross value added in unregistered manufacturing in 1988-89 at 1980-81 prices	GROSS VALUE ADDED (Rs. crores)									Annual growth rate % (Semi-log) basis
		1980-81	1981-82	1982-83	1983-84	1984-85	1985-86	1986-87	1987-88	1988-89	
1.	2.	3.	4.	5.	6.	7.	8.	9.	10.	11.	12.
Unregistered Manufacturing											
food products	4.90	521	563	527	610	596	621	632	621	740	3.53
beverages, tobacco, etc.	2.34	332	365	371	344	365	335	361	297	354	-0.74
cotton textiles											
wool, silk, etc.	23.78	2844	2956	3071	3042	3322	3407	3475	3593	3594	3.19
jute textiles											
textiles products											
wood, furniture, etc.	4.53	918	940	844	905	763	844	800	707	684	-3.61
paper & printing etc.	2.53	214	209	217	261	294	323	364	371	382	9.21
leather & fur products etc.	1.91	233	247	246	258	269	264	270	275	288	2.32

1.	2.	3.	4.	5.	6.	7.	8.	9.	10.	11.	12.
rubber, petroleum etc.	1.48	100	152	130	160	195	175	198	206	223	8.78
chemicals etc.	2.54	178	193	204	208	235	265	289	331	384	9.84
non-metalic products	2.99	293	331	337	357	339	342	393	387	452	4.20
basic metal industries	1.25	131	110	114	104	140	140	165	177	189	6.77
metal products	9.09	778	866	1022	1077	1080	997	1082	1333	1373	6.19
non-electrical machinery tools and parts	4.68	439	487	491	525	560	571	622	611	707	5.38
electrical machinery	4.53	198	205	229	282	294	397	503	662	684	18.91
transport machinery	5.96	526	568	585	649	692	714	762	799	900	6.53
other manufacturing	9.83	486	725	752	508	596	741	1143	1321	1486	13.29
repairing services	17.68	1312	1421	1477	1679	1637	2028	2151	2469	2672	9.50
Total	100.00	9503	10338	10617	10969	11377	12164	13210	14160	15112	5.72

Source: CSO: National Accounts Statistics.

Table XII

Size Distribution of Factories in 1986-87

			Percentage distribution of		
			Employment	Productive Capital	Value Added
		1.	2.	3.	4.
I.	Employment size range				
	0	- 49	17.3	6.3	9.5
	50	- 99	10.6	4.9	5.8
	100	- 199	10.1	5.8	7.8
	200	- 499	17.6	21.9	18.6
	500	- 999	14.6	24.8	17.9
	1000	- 1999	11.7	19.4	18.8
	2000	- 4999	11.7	9.2	13.6
	5000	and above	6.4	7.7	8.0
		Total	100.0	100.0	100.0
II.	Range of gross value of plant and machinery (Rs. million)				
	up to	0.10	10.25	0.93	2.76
	0.10	- 0.25	5.63	1.05	2.30
	0.25	- 0.50	5.24	1.21	2.35
	0.50	- 0.75	3.09	0.78	1.86
	0.75	- 1.00	2.30	0.64	1.16
	1.00	- 2.00	5.52	1.97	3.71
	2.00	- 2.50	1.83	0.76	1.44
	2.50	- 3.50	2.49	1.03	1.76
	above	3.50	61.55	91.32	81.86
	unspecified		2.10	0.31	0.80
		Total	100.00	100.00	100.00

Source: Annual Survey of Industries, 1986-87 Summary Results for Factory Sector, CSO

Table XIII

	Average Annual growth in industry		Production elacticities of Employment
	Production	Employment	
1951-56	7.27	5.16	0.71
1956-61	13.33	8.50	0.64
1976-79	4.71	2.86	0.61
1980-87	7.55	1.87	0.25

Source: Pradhan Prasad E.P.W., Jan., 27, 1990)

Table XIV

Budget Deficit of Central Government

	Net Revenue Expenditure	Net Capital Expenditure	Total Revenue Expenditure -External Grants.	Budget Deficit (Col 1 +2-3	Budget Deficit as % of GDP
	(1)	(2)	(3)	(4)	(5)
1980-81	14347	5550	11937	7960	(5.9)
1981-82	15333	7392	14631	8094	(5.1)
1982-83	18621	8215	17036	9800	(5.5)
1983-84	22072	8957	19366	11663	(5.6)
1984-85	27459	11260	22970	15749	(6.8)
1985-86	33213	13384	27112	19485	(7.4)
1986-87	40247	15516	32462	23301	(7.9)
1987-88	45841	14810	36545	24106	(7.2)
1988-89	53786	16196	42991	26991	(6.9)
1989-90	64205	18096	51341	30960	(7.1)

Table XV

Revenue Surplus/Deficit, All-India, Since 1978-79

(Rs crore)

Year	All-India Centre, States & UTs	Centre
1978-79	+ 1428	+ 292
1979-80	+ 854	- 694
1980-81	+ 124	- 776
1981-82	+ 1017	- 293
1982-83	- 366	- 1254
1983-84	- 2180	- 2398
1984-85	- 4396	- 3498
1985-86	- 5020	- 5565
1986-87	- 7755	- 7776
1987-88	-10379	- 9137
1988-89	-13099 (RE)	-11030 (RE)
1989-90	-10466 (RE)	- 7012 (RE)
		Σ 40,520

Source: Ministry of Finance, Indian Economic Statistics (Public Finance) and Central Budgets

Table XVI

Tax Revenue, Internal, External Borrowings and Debt Service (Ammortisation & Interest Payment) of the Centre

(Rs. Crores)

Items	1971-72 (Actuals)	1981-82 (Actuals)	1985-86 (Actuals)	1989-90 (Actuals)	1990-91 (R.E.)
1. Tax Revenue					
Gross Collection	3872	15816	28670	51636	58916
Net-Centre's Share	2928	11542	21140	38349	44318
2. Internal Market Borrowings					
Gross	632	3198	5543	8044	8988
Less: Repayments	334	285	659	640	988
Net	298	2913	4884	7404	8000
3. Cumulative Internal Debt[1] at the beginning of	* 14043	* 29008	58537	114498	133193
4. External Assistance					
Gross	540	1451	2145	4443	6241
Less: Repayments	194	422	696	1848	2257
Net	346	1029	1449	2595	3984
5. Cumulative Total Liabilities at the beginning of	* 18836	* 58723	113441	229771	268192
6. Interest Payments (in respect of all Liabilities)	670	3195	7512	17735	21850[2]
7. Average Interest Paid	3.6%	5.4%	6.6%	7.7%	8.1%

*Relates to R.E.
1) Includes market loans, treasury bills, special securities, compensation, and other bonds, special *bearer bonds & securities issued to Internal Financial Institutions.
2) Break-up is as follows:
   1. Internal Debt   9792
      i) Market Loans   6371
      ii) Discount on Treasury Bills   1440
      iii) others   1981
   2. External   1810
   3. Small Savings, Deposit Certificates etc.   4121
   4. Provident Fund   5686
   5. Reserve Funds etc.   441

Source: Central Budget Papers.

Table XVII
Centre's Revenue Receipts and Revenue Expenditure
(As per cent of GDP)

	1980-81	1982-83	1983-84	1984-85	1985-86	1986-87	1987-88	1988-89	1989-90	1990-91 (R.E.)
1. Tax Revenue (Net of States's Share)	6.9	7.3	7.5	7.6	8.1	8.3	8.4	8.6	8.7	8.6
2. Non-Tax Revenue	2.5	2.8	2.4	2.9	3.1	3.4	3.1	2.8	3.5	2.9
3. Total Current Revenue (1+2)	9.4	10.2	9.9	10.5	11.2	11.7	11.5	11.4	12.2	11.5
4. Total Current Expenditure	10.0	10.9	11.0	12.0	13.3	14.4	14.3	14.1	14.9	14.9
(a) Interest Payments	2.0	2.2	2.3	2.6	2.9	3.2	3.4	3.6	4.0	4.3
(b) Subsidies[1]	1.2	1.3	1.3	1.6	1.8	1.9	1.8	2.0	2.5	1.8
(c) Defence Expenditure[2]	2.6	2.7	2.7	2.8	2.9	3.4	3.0	2.9	2.7	2.4
(d) Grants to States & UTs	2.1	2.0	2.1	2.3	2.7	2.7	2.8	2.6	2.0	2.2
(e) Others	2.2	2.6	2.5	2.8	3.0	3.3	3.3	3.0	3.8	4.2
5. Revenue Account Surplus (+)/ Deficit (—) (3—4)	—0.6	—0.7	—1.2	—1.5	—2.1	—2.7	—2.7	—2.7	—2.7	—3.4

1. Excludes subsidies on imported fertilizer up to 1985-86.
2. Includes defence pension.

Note:—(a) The ratio in this Table from 1980-81 onwards are based on the new series on GDP released by the CSO and therefore these will differ from the figures given in the earlier issues of Economic Survey.
(b) Totals may not add up because of rounding.

Table XVIII

Growth of Liabilities of Government of India: 1980-81 to 1990-91

(Rs. in crores)

Year	Internal Debt	Small savings	Provident Funds & other Accounts	Reserve Funds & Deposits	Total internal Liabilities	External Liabilities	Total Liabilities
Sixth Plan							
1980-81	30864	7976	5977	3634	48451	11298	59749
	(22.7)	(5.9)	(4.4)	(2.7)	(35.7)	(8.3)	(44.0)
1981-82	35653	9375	7203	3627	55858	12328	68186
	(22.4)	(5.9)	(4.5)	(2.3)	(35.0)	(7.7)	(42.7)
1982-83	46939	11098	8789	4364	71190	13682	84872
	(26.4)	(6.2)	(5.0)	(2.5)	(40.1)	(7.7)	(47.8)
1983-84	50263	13506	10368	6004	80141	15120	95261
	(24.3)	(6.5)	(5.0)	(2.9)	(38.7)	(7.3)	(46.0)
1984-85	58537	17157	12547	8567	96804	16637	113441
	(25.4)	(7.5)	(5.4)	(3.7)	(42.0)	(7.2)	(49.2)
Seventh Plan							
1985-86	71039	21449	15410	11433	119331	18153	137484
	(27.0)	(8.2)	(5.9)	(4.4)	(45.5)	(6.9)	(52.4)
1986-87	86312	24725	20204	15006	146247	20299	166546
	(29.4)	(8.4)	(6.9)	(5.1)	(49.8)	(6.9)	(56.7)
1987-88	98646	28358	26170	19164	172338	23223	195561
	(29.7)	(8.5)	(7.9)	(5.7)	(51.8)	(7.0)	(58.8)
1988-89	114498	33833	34702	20992	204025	25746	229771
	(29.3)	(8.6)	(8.9)	(5.3)	52.1)	(6.6)	(58.7)
1989-90 (Revised Estimate)	133361	40583	43643	20809	238396	28517	266913
	(30.1)	(9.2)	(9.9)	(4.7)	(53.9)	(6.4)	(60.3)
Eighth Plan							
1990-91	151037	45583	54590	23815	275025	31851	306876

(Budget Estimates)
Note: Figures in brackets represent percentages to Gross Domestic Product at current market prices.
Sources: RBI Annual Report, 1989-90.

Table XIX

Number of Returns, Income and Tax Payable, All Status,* 1981-82 and 1988-89

Range of returned income ('000 Rupees)	1981-82 Returns Number in Thousands	1981-82 Returns % to the total	1981-82 Income returned* Rs. Crores	1981-82 Income returned* % to the total	1981-82 Tax Payable Rs. Crores	1981-82 Tax Payable % to the Total	1988-89 Returns Number in Thousands	1988-89 Returns % to the total	1988-89 Income returned** Rs. Crores	1988-89 Income returned** % to the total	1988-89 Tax Payable@ Rs. Crores	1988-89 Tax Payable@ % to the total
T.L - 20	914.7	65.05	1255.0	24.93	91.0	6.64	864.8	23.69	1600.7	7.52	25.5	0.47
20 - 50	371.4	26.41	1081.3	21.48	164.6	12.01	1726.1	47.28	5537.4	26.01	616.5	11.39
50 - 100	87.2	6.20	583.9	11.60	132.1	9.64	888.6	24.34	6395.7	30.04	1341.7	24.79
100 - 200	22.5	1.60	302.2	6.00	82.2	6.00	122.9	3.37	1651.6	7.76	480.6	8.88
200 - 300	4.5	0.32	106.8	2.12	37.6	2.74	24.0	0.66	585.6	2.75	225.1	4.16
300 - 400	1.8	0.13	62.4	1.24	24.3	1.77	10.1	0.28	348.3	1.64	139.0	2.57
400 - 500	0.9	0.06	40.1	0.80	16.5	1.20	6.4	0.18	286.5	1.35	112.7	2.08
above 500	3.1	0.22	1603.2	31.84	822.3	60.00	8.0	0.22	4883.7	22.94	2470.8	45.65
Total	1406.1	100.00	5034.9	100.00	1370.6	100.00	3650.9	100.00	21289.5	100.00	5411.9	100.0

*Individuals, Hindu Undivided Families, Registered Firms & Others, and Companies.
**After B/F loss, etc. set off and deductions under Chapter VI A.
@Including surcharge/surtax

TL= Taxable Level.

Source: All India Income Tax Statistics, Assessment year, 1981-82 and 1988-89, Directorate of Income Tax, New Delhi.

Table XX
Merchandise Trade and Balance of Payments as percent of GDP since 1980-81

(Rs. Crores)

Years	Merchan-dise Imports	Merchan-dise Exports (DGCI&S)	Balance of Trade (DGCI&S)	Balance of Trade Payments (RBI)	% to GDP Merchan-dise imports	% to GDP Merchan-dise exports (DGCI&S)	% to GDP Balance of trade	Balance of Payments Trade	Balance of Payments Payments	Memo: GDP at Market Prices
1980-81	12549	6711	-5838	-5967	-1657	9.24	4.94	-4.30	-4.39	-1.22 (1.5)	135812
1981-82	13608	7806	-5802	-6121	-2317	8.54	4.90	-3.64	-3.84	-1.45 (1.6)	159420
1982-83	14293	8803	-5490	-5776	-2296	8.05	4.96	-3.09	-3.25	-1.29 (1.4)	177588
1983-84	15831	9771	-6060	-5871	-2262	7.65	4.72	-2.93	-2.04	-1.09 (1.3)	206857
1984-85	17134	11744	-5390	-6721	-2852	7.43	5.09	-2.34	-2.91	-1.24 (1.4)	230679
1985-86	19658	10895	-8763	-9586	-5927	7.49	4.15	-3.34	-3.65	-2.26 (2.4)	262603
1986-87	20096	12452	-7644	-9354	-5830	6.85	4.24	-2.61	-3.19	-1.99 (2.2)	293361
1987-88	22244	15674	-6570	-9296	-6293	6.69	4.71	-1.98	-2.00	-1.89 (2.1)	332553
1988-89	28194	20302	-7892	-14003	-10410	7.14	5.14	-2.00	-3.55	-2.64 (2.8)	394992
1989-90	35412	27681	-7731	…-10183 E		8.00	6.25	-1.75	…	-2.30 E (2.4)	442769 E
1990-91	42000	32500	-10500	…	…	8.30	6.44	0.00	…	-2.90 E (3.0)E	504760 E

Parentheses show the C.S.O. estimates of foreign savings.   E-Estimate

Table XXI
Trade Merchandise

(Rs. Crores)

Years	Private			Government/Public			Total		
	Credit	Debit	Net	Credit*	Debit	Net	Credit	Debit	Net
1.	2.	3.	4.	5.	6.	7.	8.	9.	10.
1970-71	1127.1	675.1	452.0	277.4	1151.0	-873.6	1404.5	1826.1	-421.6
1980-81	5188.4	4735.3	453.1	1388.0	7808.3	-6420.3	6576.4	12543.6	-5967.2
1985-86	9025.3	9641.7	-616.4	2552.3	11521.9	-8969.6	11577.6	21163.6	-9586.0
1986-87	10813.8	11396.7	-582.9	2501.2	11272.2	-8771.0	13315.0	22668.9	-9353.9
1987-88	13648.5	12643.0	1005.5	2747.9	13049.5	-10301.6	16396.4	25692.5	-9296.1
1988-89	17558.3	17447.2	111.1	3088.4	16755.1	-13666.7	20646.7	34202.3	-13555.6

In view of the amounts shown as zero in RBI reports, figures have been taken from Public Enterprises Survey—1989-90; and include canalised and non-canalised goods; these are deducted from credit in col. 2.

*

Table XXII

Key Indicators of India's Balance of Payments
(As per cent of GDP)

Year	Exports	Imports	Net Invisibles	Trade Balance	Current Account Balance
1	2	3	4	5	6
1980-81	4.8	9.2	3.2	-4.4	-1.2
1981-82	4.9	8.7	2.4	-3.8	-1.5
1982-83	5.1	8.4	2.0	-3.2	-1.3
1983-84	4.9	7.7	1.7	-2.8	-1.1
1984-85	5.2	8.1	1.7	-2.9	-1.2
Average 1980-85	5.0	8.4	2.2	-3.4	-1.3
1985-86	4.4	8.1	1.4	-3.7	-2.3
1986-87	4.5	7.7	1.2	-3.2	-2.0
1987-88	4.9	7.7	0.9	-2.8	-1.9
1988-89	5.3	8.9	0.8	-3.5	-2.7
1989-90	6.4	9.3	0.6	-2.9	-2.3
Average 1985-90	5.1	8.3	1.0	-3.2	-2.2

Note: The ratios have been computed on the basis of the country's balance of payments data as given in appendix table 6.2 and the CSO estimates of GDP at current market prices. Official grant receipts and US embassy expenditure in India out of PL-480 Rupee Fund are taken as current account receipts in conformity with the balance of payments statistics publised by the RBI.

Table XXIII

Export Performance during Sixth and Seventh Plans

		Sixth Plan (1980-1985)	Seventh Plan (1985-1990)
(a)	Exports as per cent of GDP	4.9	5.0
(b)	Value growth (Per cent per annum)		
	In Rupees	13.0	19.8
	In Dollars	4.5	11.6
	In SDRs	10.0	6.6
(c)	Volume growth (Per cent per annum)		
	Target	9.0	7.0
	Actuals	2.7	6.3*
(d)	Level of exports (Annual average)		
	In Rupee crores	8,967	17,387
	In Dollars million	9,125	12,267
	In SDRs million	8,251	9,821

*Growth ratio for the first four years over Seventh Plan for which DGCI & S Data available.

Table XXIV

Import Trends during Sixth and Seventh Plans

		Sixth Plan (1980-1985)	Seventh Plan (1985-1990)
(a)	Imports as per cent of GDP	8.1	7.3
(b)	Value growth (Per cent per annum)		
	In Rupees	13.9	16.0
	In Dollars	6.2	8.2
	In SDRs	11.4	3.4
(c)	Volume growth (Per cent per annum)		
	Target	9.5	5.8
	Actuals	6.9	9.8*
(d)	Level of imports (Annual average)		
	In Rupees crores	14,683	25,129
	In Dollars million	15,110	17,943
	In SDRs million	13,571	14,489

*Average growth in the first four years of Seventh Plan for which DGCI&S data are available.

Table XXV

External Commercial Borrowings*

(Rs crores)

	1985-86	1986-87	1987-88	1988-89	1989-90	1990-91(P)
I. Authorisations	1700	1396	2654	4314	5479	3414
II. Gross disbursements	1799	2474	2252	4069	4196	3050
III. Debt service payments	1175	1565	1736	2224	3041	4006
(a) Amortisation	565	796	871	1103	1455	2137
(b) Interest payments	610	769	865	1121	1586	1869
IV. Net capital inflow [2—3(a)]	1234	1678	1381	2966	2741	913

P—Provisional
*Excludes borrowings upto 1 year maturity. The estimates are based on data base of the ECB Division of the Dept. of Economic Affairs, Ministry of Finance.

Table XXVI

Year	External debt in million dollars	Debt Servicing/ Exports (Per cent)	Debt/GNP (Per cent)
1980	20,561	9.2	11.9
1981	22,567	10.4	—
1982	27,376	13.8	14.9
1983	31,891	16.6	16.0
1984	33,857	18.1	17.6
1985	40,886	22.3	19.2
1986	48,351	31.6	21.7
1987	55,325	29.0	21.7
1988	57,513	28.9	21.4
1989	69,700	26.3	23.9
1990	80,000 (rough estimates)		

Source: World Debt Tables

Table XXVII

Norms for Indebtedness and Figures for India as given in the World Debt Tables

Criteria	Severely Indebted Country	Moderately Indebted Country	India (1989)
Debt/GNP	50%	30-50%	23.9%
Debt/Exp	275%	165-275%	258.4%
Debt Servicing Exp	30%	18-30%	26.3%
Accrued Int./Exp.	20%	12-20%	14.2%

Table XXVIII

Imports of Principal Commodities

(Rs crore)

	1985-86	1986-87	1987-88	1988-89 (Partially Revised)	1989-90 (Provisional)
(1) Bulk imports *of which*	10,763	8,024	9,084	11,175	14,239
(a) POL	4,989	2,811	4,043	4,374	6,274
(b) Fertilisers	1,436	921	508	928	1,776
(c) Non-ferrous metals	542	517	639	786	1,253
(d) Iron and steel	1,395	1,556	1,320	1,937	2,305
(e) Bulk consumption goods (cereals, edible oils, etc)	1,466	1,179	1,503	1,741	914
(2) Non-bulk imports *of which*	8,895	12,072	13,160	17,019	21,173
(a) Capital goods	4,286	6,488	6,566	6,939	8,831
(b) Export related items	2,366	2,856	3,351	5,463	6,803
(c) Others	2,243	2,728	3,243	4,617	5,539
Total	19,658	20,096	22,244	28,194	35,412

Source: EPW Feb 1991.

Table XXIX

Exports of Principal Commodities

(Rs crore)

	1985-86	1986-87	1987-88	1988-89 (Partially Revised)	1989-90 (Provisional)
(1) Primary products	3803	4139	4160	4374	5951
of which					
(a) Agricultural and allied products	3018	3422	3411	3347	4571
(b) Ores and minerals	785	717	749	1027	1380
(2) Manufactured goods	6647	7902	10865	14641	20310
of which					
(a) Cotton yarn fabrics, made-ups	574	637	1155	1131	1480
(b) Readymade garments	1067	1331	1813	2097	3224
(c) Leather and leather manufacturers	770	922	1251	1490	1951
(d) Gems and jewellery	1503	2074	2617	4399	5296
(e) Chemicals and allied products	498	583	801	1534	2158
(f) Engineering goods	954	1133	1502	2364	3321
Total (including others)	10895	12452	15674	20302	27681

Economic and Political Weekly February 9, 1991

Table XXX

Pattern of External Liabilities

(Rs crore)

End of March	External Assistance	External Commercial Borrowing	Liabilities to IMF	NRI Deposits	Total	Col. 3 + Col 5 as PC of Total
(1)	(2)	(3)	(4)	(5)	(6)	(7)
1985	24,004	6,908	4,888	3,819	39,619	27.1
1986	26,638	8,075	5,271	6,650	45,634	32.3
1987	32,312	11,243	5,548	7,847	56,950	33.5
1988	36,578	13,543	4,732	10,054	64,907	36.4
1989	46,838	19,147	3,696	14,154	83,835	39.7
1990	54,095	24,500	2,573	17,831	98,999	42.7

Economic and Political Weekly February 9, 1991

Table XXXI

Projected and Actual Resource Generation** by the Public Sector During the Plans

(Rs Crores at base level prices)

Plan	Projections	Actuals
First Plan (1951-56)*	170	115
Second Plan (1956-61)*	150	167
Third Plan (1961-66)	500	435
Fourth Plan (1969-74)	2029	1145
Fifth Plan (1974-79)	849	N.A.
Sixth Plan (1980-85)	9395	5810
Seventh Plan (1985-90)	35495	13246

*Railways only.   **Gross
Source: Five Year Plans and Expenditure Budgets of the Central Government.

Table XXXII

Share of Public Sector in Gross Domestic Capital Formation

(Rs crore)

Year	Aggregate*	Public Sector	Admn Department	Deptt. Enter-prises (DEs)	Non-Deptt Enter-prises (NDEs)	Total (5)+(6)
1	2	3	4	5	6	7
1980-81	31016 (30867)	14000 (45.1)	3101 (22.2)	3565 (25.4)	7334 (52.4)	10899 (35.1)
1984-85	51653 (46440)	26493 (51.3)	5434 (20.5)	6068 (22.9)	14221 (56.6)	21059 (40.8)
1985-86	66729 (59917)	30875 (46.3)	6728 (21.8)	6796 (22.0)	17351 (56.2)	24147 (36.2)
1986-87	71308 (63646)	35422 (49.7)	7704 (21.7)	6816 (19.2)	20902 (59.0)	27718 (38.9)
1987-88	76140 (72134)	35404 (46.5)	8135 (23.0)	7175 (20.3)	20094 (56.7)	27269 (35.8)
1988-89	95380 (93544)	41125 (43.1)	—	—	—	—

Note: *Unadjusted for errors and omissions; adjusted in parentheses.
Parentheses in col 3 and 7 show percentage share in (2), parentheses in col 4 to 6 show percentage share in (3).
Source: Central Statistical Organisation, National Accounts Statistics, 1990, pp. 57, 93 and 97

Table XXXIII

Gross Domestic Savings: Share of Public Sector

(Rs crore)

Year	Aggregate*	Public Sector	Admn Department	Deptt. Enterprises (DEs)	Non-Deptt Enterprises (NDEs)	Total (5)+(6)
1	2	3	4	5	6	7
1980-81	28773	4654 (16.2)	2559 (55.0)	245 (5.3)	1850 (39.7)	2095 (7.3)
1984-85	40438	6526 (16.1)	-315 (-4.8)	732 (11.2)	6109 (93.6)	6841 (17.0)
1985-86	53683	8457 (15.8)	-474 (-5.6)	1419 (16.8)	7512 (88.8)	8931 (16.6)
1986-87	57221	7981 (13.9)	-2400 (-30.0)	1494 (18.7)	8887 (111.4)	10381 (18.1)
1987-88	65309	6861 (10.5)	-5773 (-84.1)	2063 (30.1)	10571 (154.0)	12634 (19.3)
1988-89	82044	6346 (7.7)	-8702 (-137.0)	2680 (42.2)	12368 (194.9)	15048 (18.3)

Parentheses in col 3 and 7 show percentage share in (2), parentheses in col 4 to 6 show percentage share in (3).

Source: Central Statistical Organisation, National Accounts Statistics, 1990, pp, 57, 93 and 97

Table XXXIV

Gross Domestic Savings: Share of Public Sector

(Rs crore)

Year	Aggregate*	Public Sector	Admn Department	Deptt. Enterprises (DEs)	Non-Deptt Enterprises (NDEs)	Total (5)+(6)
1	2	3	4	5	6	7
1980-81	12087	4895 (40.4)	764 (15.8)	1461 (29.7)	2670 (54.5)	4131 (34.2)
1984-85	22091	9408 (42.6)	1578 (16.8)	2664 (28.3)	5166 (54.9)	7830 (35.4)
1985-86	26239	11388 (43.4)	1940 (17.0)	3190 (28.0)	6258 (55.0)	9448 (36.0)
1986-87	29929	13095 (43.7)	2264 (17.3)	3474 (26.5)	7357 (56.2)	10831 (36.2)
1987-88	33876	14942 (44.1)	2627 (17.6)	3845 (25.7)	8470 (56.7)	12315 (36.3)
1988-89	39455	17567 (44.5)	—	—	—	—

Parentheses in col 3 and 7 show percentage share in (2), parentheses in col 4 to 6 show percentage share in (3).

Source: Central Statistical Organisation, National Accounts Statistics, 1990, pp, 57, 93 and 97

Table XXXV

Employment and Productivity Indices of Central Public Sector Enterprises (for the period between 1973-74 and 1988-89)

Year	Labour Productivity	Capital Productivity	TFPG Solow	TFPG Kendrik	TFPG Translog	Employment generation (lakh number of persons)	Growth of employment general (%)
1973-74	58.83	20.95	—	—	—	13.44	—
1974-75	66.24	26.59	0.1611	0.2249	0.1534	14.32	6.55
1975-76	75.47	21.55	0.0533	-0.1712	0.0248	15.04	5.03
1976-77	87.76	21.37	0.0881	0.0097	0.0933	15.75	4.72
1977-78	86.49	19.23	0.0444	0.0950	0.0456	16.38	4.00
1978-79	97.49	19.71	0.0861	0.0357	0.0887	17.03	3.97
1979-80	88.52	18.86	0.0808	0.0500	0.0819	17.75	4.23
1980-81	85.08	18.97	0.0279	0.0007	0.0248	18.39	3.61
1981-82	108.61	23.20	0.2272	0.2061	0.2255	19.39	5.44
1982-83	126.86	23.21	0.0819	0.0181	0.0834	20.24	4.38
1983-84	131.76	22.21	0.0005	0.0348	0.0008	20.72	2.37
1984-85	140.25	21.11	0.0116	0.0378	0.0103	21.07	1.69
1985-86	143.02	19.42	0.0291	0.0717	0.0303	21.54	2.23
1986-87	157.58	19.30	0.0461	0.0059	0.0465	22.11	2.65
1987-88	159.41	17.40	0.0444	0.0905	0.0450	22.14	0.14
1988-89	181.16	18.02	0.0819	0.0459	0.1038	22.09	0.23

Source: A statistical review of Central Government Enterprises: 1988-89, CMIE, Bombay, March 1990.

Table XXXVI

Poverty in 2000, by Developing Region

Region	Incidence of Poverty		Number of poor (millions)	
	1985	2000	1985	2000
Sub-Saharan Africa	46.8	43.1	180	265
East Asia	20.4	4.0	280	70
China	20.0	2.9	210	35
South Asia	50.9	26.0	525	365
India	55.0	25.4	420	255
Eastern Europe	7.8	7.9	5	5
Middle East, North Africa, and other Europe	31.0	22.6	60	60
Latin America and the Caribbean	19.1	11.4	75	60
Total	32.7	18.0	1,125	825

Note: The incidence of poverty is the share of the population below the poverty line, which is set at $370 annual income (the higher line used in the Report).

Source: For 1985, Table 2.1; for 2000, World Bank estimates.

Index

Abid Hussain Report, 57
Advanced Developed Countries (ADCs), 58-59
Aggarwal, 75
Ambedkar, 4
Arunachalam, M.V., 94
Balakrishna, Pulapre, 46
Balance of Payments (BoPs)
  cirsis, 38-43, 56
  factors for, 41-43
  government strategies, 55-56
  foreign exchange and, 36-38
  Government statement on 43-46
  Key indicators of, 225
  merchandise trade and, 223
  vicious circle, 38-41
Bardan, Pranob, 67, 112
Birla, Aditya, 102
Black money, 150
Bofors deal, 88
Borrowings, tax revenue and debt, 219
Boyd, 44
Brown, Lester, 126
Budget deficits, 199, 217
Bureaucracy, 114

Centre expenditure, 197
Chakravarty, Sukhomoy, 23, 65, 112
Community participation, concept of, 143-44
Corruption, 150
Credit Rating and Information Services of India Ltd. (CRISIC), 45-46
Das, Purshottam Dass Thakur, 5
Das Gupta, 76
Debt, tax revenue and borrowings, 219
Democratic revolution, 3
Demography, effects on economic development, 138-46
Development Economies, 126
Development model, 11-16
Direct Taxes Enquiry Committee, 27
Domar, E., 15, 122
Domestic savings and investment, 206
Economic Commission for Latin America, 121
Egalitarian revolution, *see*, Socialist revolution

Election Commission, 8
Employment,
    and productivity indices, 235
    models of, 119-21, 132-33
Export,
    commodities, 55-56
    Performance in sixth and seventh plan, 226
    Promotion constraints, 54-57
    strategy, 55
    substitution, 47-54
External commercial borrowings, 229
External debt on India, 44-45
European Commission (EC) during 1992, 58
FERA laws, 91, 94, 158-59
Factories, size distribution of, 216-17
Family planning programme, 144-46
Fei, 120-21
Fel'dman's model, 15
Fical policy, 26-35
Francis, Scott, 139
Friedman, 11
GIC, 79
Gadgil, D.R., 116
Gandhi, Indira, 27, 103, 107, 114, 116, 157
Gandhi, M.K., 1-3, 90, 119, 137, 149
Gandhi, Rajiv, 40, 109, 114
German Development Institute, 58
Ghosh, Arun, 38, 40
Goenka, R.P., 93
Government expenditure and public sector investment, 18
Gross domestic capital formation, 208, 232
Gross domestic savings, 208, 233-34
Gross national product, 205

Gross rates in value added prices, by industry, 212-15
Growth theory, critique of, 121-37
Hansen, Alvin, 34
Harrod-Domer model, 122
Hashmi, S.R., 207
Hirishi, 15
Hirschman, 126-27
Hussain, Abid, 57
Hussain, Akmal, 32
ICICI, 45
ICOR, 23-24, 33
IMF, 39, 42, 57, 95, 100-01, 148
Imperial Bureaucratic State (IBS), 2-6, 8
Import,
    effects of, 51
    of technologies, 51-54
    protection to, 49-51
    substitution, 47-54
    trends in sixth and seventh plan, 227
Income, tax payable and returns, number of, 222
Incremental Gross Capital output ratio, 207
Indebtedness, norms of, 229
India,
    annual budget, 87-89
    income tax payers, 74-76
    industrial sixkness, 80-82
    middle class in, 72-82
    new middle class in, 74-76
    rentier class in, 83-90
    tax system in, 74-76, 87
India's economic policy,
    crises in, 9-16, 101-14
    demographic features and, 138-46
    development model, 11-16
    external debt and, 44-45
    features of, 54
    financial variables, 32

Index

fiscal policy in, 26-35
growth rate and, 115-37
impact of, 147-60
indicators of, 198
Mahalanobis Plan, 5-7, 9, 113, 134
monetary variables and, 31-35
planning strategies, 105-06
Smie 1947-48, 1-8
structural issues in, 101-14
tax evasion by professionals, 28-30
vested interest, 116-19
vicious circles, 17-25
    cost of growth, 24-25
    critical aspect, 30-31
    lacunae in plan, 21-24
    poverty, 19-21
Industrial-technological revolution, 3
Investment and domestic savings, 206
Jadhav, Narender, 31
Kalecki, 12, 82
Kanwar, Onkar S., 94
Keynes, J.M., 1, 11, 30, 34, 117, 120
Krucger, A.O., 83
Kuznet, 124
Larner Aba P., 34
Leontief, 123
Less Developed Countries (LDCs), 58-59, 116, 119, 126, 129, 131
Lewis, Arther, 12, 120-21, 124
Liabilities of Government of India, growth of, 221
Liberalisation, 91-100
    conditions for, 97-99
    domestic, 92-97
    external, 92, 97
    internal, 92
MRTP Laws, 94

MacNmara, 143
Mahalanobis Plan, 5-7, 9, 12, 15, 21-25
Malthus, 1
Market failures, and public sector, 62-65
Marx, Karl, 1, 11, 82, 105, 117, 151, 154
Mendez, Jorge, 119, 121
Merchandise trade and balance of payments, 223
Middle class in India, 72-80
    concept of, 72-74
    consumption pattern, 77-80
    income tax payers, 74-76
    industrial sickness and, 80-82
    new, 74-77
Minhas, B.S., 63
Mitra, 24
Modi, K.N., 94
Moody, 45
Morris, Sebastian, 70
NRI Funds, 79
Nanda, Rajan, 93
Narasimha Rao, P.V., 148, 160
National Development Council, 142
Nationalist revolution, 3
Nayak, 75
Nayyar, Deepak, 112
Nehru, J.L., 2-5, 9, 81, 118-19
Nehru-Mahalanobis Plan, 17, 26, 37, 52, 57, 61-62, 65, 81, 95, 101, 103, 106-08, 111, 113
    lacunae in, 21-24
Net domestic product, trends in, 209
New industrial policy, 158-60
Nurkse, 12
PL 480, 21
Planning Commission, 3-5, 15, 21, 32-33, 38, 41-42, 66,

108, 116, 119, 136
Political economy of India, 1-8
Population,
   by expenditure group, 211
   macro economic aggregates and, 200-03
Population Commission, 138
Poverty, 236
Prebisch, Paul, 121, 124
Pricing policy, 49-50
Principal commodities,
   export of, 229
   import of, 230
Private sector, and public sector, 71
Professional class, tax evasion by, 28-30
Public Accounts Committee, 80
Public sector,
   facts of, 66-67
   gross domestic savings and, 233-34
   in China, 67
   investments in, 66
   market failures and, 62-65
   parasitic, 68-71
   private sector and, 71
   resource generation by, 231
   share in gross domestic capital formation, 232
   tasks, 61-62
Public sector investment, and government expenditure 18
Puri, Ashwini Kumar, 93
Raipuria, Kalyan, 66
Ramati, Yohanar, 120
Rani's, 120-21
Real domestic product, growth of, 204
Rentier class, 83-90
   creation of, 89-90

Resource generation, by public sector, 231
Returns, income and tax payable number of, 222
Revenue capital receipts, 197
Revenue receipts and expenditure, 220
Revenue surplus, 218
Ricardo, David, 1, 11
Rishi, 44
Robinson, Joan, 11, 82
Roy, 81
Rupee, devaluation of, 40-41
Santhanam Committee, 88
Schumpter, 1
Sen, A.K., 81, 127, 134
Shetty, S.L., 77-78
Singer, 124
Singh, Balwant, 31
Singh, Manmohan, 160
Sinha, Yashwant, 43
Smith, Adam, 1, 11, 117
Socialist revolution, 3
South Asian Child, 126-28
Srafa, 128
Stalin, 21
Tax payable, returns and income, number of, 222
Tax revenue, borrowing and debt, 219
Trade merchandise, 223-34
Trade prospects, 58-60
UNICEF, 126
UTI, 79
United Nations Population Fund Seminar, 140
Workers, distribution of, by sectors, 210
World Bank, 57, 95, 100-01, 143, 149
World debt table, India in, 229